Series / Number 07-022

APPLIED
REGRESSION

An Introduction

MICHAEL S. LEWIS-BECK
University of Iowa

 SAGE PUBLICATIONS / Beverly Hills / London

Copyright © 1980 by Sage Publications, Inc.

Printed in the United States of America

For information address:

SAGE Publications, Inc.
275 South Beverly Drive
Beverly Hills, California 90212

SAGE Publications Ltd
28 Banner Street
London EC1Y 8QE, England

International Standard Book Number 0-8039-1494-6

Library of Congress Catalog Card No. L.C. 80-5821

FIFTH PRINTING, 1983

When citing a professional paper, please use the proper form. Remember to cite the
correct Sage University Paper series title and include the paper number. One of the
two following formats can be adapted (depending on the style manual used):

(1) IVERSEN, GUDMUND R. and NORPOTH, HELMUT (1976) "Analysis of Variance." Sage University Paper series on Quantitative Applications in the Social
Sciences, 07-001. Beverly Hills and London: Sage Pubns.

OR

(2) Iversen, Gudmund R. and Norpoth, Helmut. 1976. *Analysis of Variance*. Sage
University Paper series on Quantitative Applications in the Social Sciences, series no.
07-001. Beverly Hills and London: Sage Publications.

CONTENTS

Editor's Introduction

It is indeed a great pleasure for us to publish, at long last, a paper on applied regression analysis. We have searched long and hard for just the right author and manuscript because of the importance of this topic to our series. In APPLIED REGRESSION: AN INTRODUCTION, Michael S. Lewis-Beck has fulfilled our hopes and expectations and then some. Dr. Lewis-Beck writes with simplicity and clarity, and without distortion. The budding social science researcher will find, I believe, that Lewis-Beck's manuscript is the ideal starting point for an introductory, nontechnical treatment of regression analysis. The emphasis here is on *applied* regression analysis, and Dr. Lewis-Beck provides several clever examples to drive home important points on the uses and abuses of regression analysis. His examples include the determinants of income, including education, seniority, sex, and partisan preference as independent variables; the factors affecting coal mining fatalities; factors affecting votes for Peron in Argentina; and several additional practical applications of regression analysis.

Lewis-Beck uses his many examples to excellent advantage in explicating the assumptions underlying regression analysis. He provides a handy list of these assumptions, and then has a lengthy section wherein he gives a very nice verbal explanation of what each assumption means in practical terms and its substantive implications. The beginner will rapidly develop an appreciation of these assumptions, their importance, and how to assess them in any particular substantive problem he or she may wish to address.

Professor Lewis-Beck provides a straightforward treatment of the slope estimate and the intercept estimate in regression analysis and their interpretation. Techniques available for assessing the "goodness of fit" of a regression line are presented and interpreted, including a discussion of the coefficient of determination and tests of significance, the latter presented within a more general discussion of confidence intervals. The discussion of significance tests, including one-tailed versus two-tailed tests, is much superior to that provided in most introductory treatments of applied regression. Another outstanding section is the treatment of the analysis of residuals, or error terms, in regression analysis. Their diagnostic potential is clearly demonstrated in the assessment of the assumptions of the regression model. This presentation is remarkably thorough, particularly given the space limitations that our series imposes on authors. We emphasize a limited, introductory treatment of these topics, and Professor Lewis-Beck is better than most of us at clarifying a complex topic in a minimum amount of space.

Finally, in the third chapter, multiple regression analysis is explicated. (The first two chapters deal only with bivariate regression. The complications of multivariate regression are presented later.) Building on the treatment of bivariate regression, the principles of multiple regression are briefly, yet lucidly covered. Each of the major points covered for bivariate regression is expanded upon in the multivariate context, and the additional complications of interaction effects and multicollinearity are given considerable attention. There is also some attention devoted to specification error and levels of measurement, including the analysis of dummy variables, in the concluding chapter. Examples abound, illustrating each of the problems covered in the text.

The importance of this paper cannot be overemphasized. Perhaps more than any other statistical technique, regression analysis cuts across the disciplinary boundaries of the social sciences. Listing its uses is not even necessary, since all social science researchers who have attempted empirical research, or who have attempted to keep abreast of recent developments in their field, have undoubtedly concluded that some understanding of regression is necessary. It is perhaps most developed in economics, where econometrics is a common part of every graduate curriculum, but researchers in psychology, sociology, political science, anthropology, mass communications, social work, public affairs, and many others are not far behind.

—John L. Sullivan, Series Editor

ACKNOWLEDGMENTS

Many people have contributed, directly or indirectly, to the development of this monograph. Foremost, I would like to thank those at the University of Michigan who began my instruction in quantitative methods: Frank Andrews, Lutz Erbring, Larry Mohr, Donald Stokes, and Herb Weisberg. Next, I wish to express my gratitude to those authors whose works have especially aided me: Harry H. Kelejian and Wallace E. Oates, Jan Kmenta, Robert S. Pindyck and Daniel L. Rubinfeld, and Edward R. Tufte. Of course, these fine teachers are in no way responsible for any errors that the manuscript may contain.

At Iowa, I would like to acknowledge Mark Williams and Meri Beth Herzberg for computer assistance, Sherry Flanagan for typing, and Carol Taylor for graphics. Last, I am grateful to the University of Iowa for awarding me an Old Gold Summer Fellowship, which allowed me to devote full-time to the preparation of the monograph.

1. BIVARIATE REGRESSION:
FITTING A STRAIGHT LINE

Social researchers often inquire about the relationship between two variables. Numerous examples come to mind. Do men participate more in politics than do women? Is the working class more liberal than the middle class? Are Democratic members of Congress bigger spenders of the taxpayer's dollar than Republicans? Are changes in the unemployment rate associated with changes in the President's popularity at the polls? These are specific instances of the common query, "What is the relationship between variable X and variable Y?" One answer comes from bivariate regression, a straightforward technique which involves fitting a line to a scatter of points.

Exact Versus Inexact Relationships

Two variables, X and Y, may be related to each other exactly or inexactly. In the physical sciences, variables frequently have an exact relationship to each other. The simplest such relationship between an *independent variable* (the "cause"), labelled X, and a *dependent variable* (the "effect"), labelled Y, is a straight line, expressed in the formula,

$$Y = a + bX,$$

where the values of the coefficients, a and b, determine, respectively, the precise height and steepness of the line. Thus, the coefficient a is referred to as the *intercept* or *constant*, and the coefficient b is referred to as the *slope*. The hypothetical data in Table 1, for example, indicate that Y is linearly related to X by the following equation,

$$Y = 5 + 2X.$$

This straight line is fitted to these data in Figure 1a. We note that for each observation on X, one and only one Y value is possible. When, for instance, X equals one, Y must equal seven. If X increases one unit in value, then Y necessarily increases by precisely two units. Hence, knowing the X score, the Y score can be perfectly predicted. A real world example with which we are all familiar is

$$Y = 32 + 9/5 \, X,$$

where temperature in Fahrenheit (Y) is an exact linear function of temperature in Celsius (X).

In contrast, relationships between variables in the social sciences are almost always inexact. The equation for a linear relationship between two social science variables would be written, more realistically, as

$$Y = a + bX + e,$$

where e represents the presence of *error*. A typical linear relationship for social science data is pictured in Figure 1b. The equation for these data happens to be the same as that for the data of Table 1, except for the addition of the error term,

$$Y = 5 + 2X + e.$$

TABLE 1
Perfect Linear Relationship between X and Y

Y = 5 + 2X	
X	Y
0	5
1	7
2	9
3	11
4	13
5	15

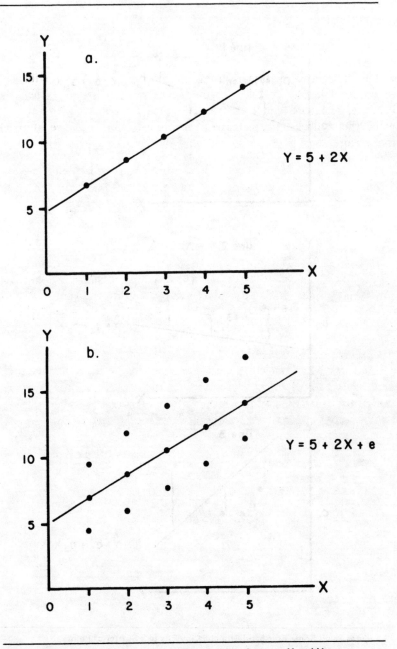

Figures 1a-b: Exact and Inexact Linear Relationships between X and Y

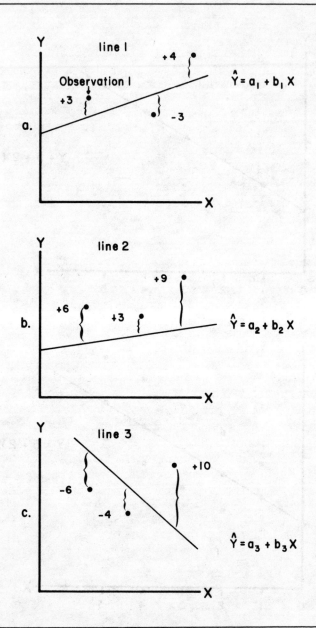

Figures 2a-c: Some Free-hand Straight Line Fits to a Scatter of Points

The error term acknowledges that the prediction equation by itself,

$$\hat{Y} = 5 + 2X,$$

does not perfectly predict Y. (The $\hat{Y}$ distinguishes the predicted Y from the observed Y.) Every Y value does not fall exactly on the line. Thus, with a given X, there may occur more than one Y. For example, with X = 1, we see there is a Y = 7, as predicted, but also there is a Y = 9. In other words, knowing X, we do not always know Y.

This inexactness is not surprising. If, for instance, X = number of elections voted in (since the last presidential election), and Y = campaign contributions (in dollars), we would not expect everyone who voted in, say, three elections to contribute exactly the same amount to campaigns. Still, we would anticipate that someone voting three times would likely contribute more than someone voting one time, and less than someone voting five times. Put another way, a person's campaign contribution is likely to be a linear function of electoral participation, plus some error, which is the situation described in Figure 1b.

The Least Squares Principle

In postulating relationships among social science variables, we commonly assume linearity. Of course, this assumption is not always correct. Its adoption, at least as a starting point, might be justified on several grounds. First, numerous relationships have been found empirically to be linear. Second, the linear specification is generally the most parsimonious. Third, our theory is often so weak that we are not at all sure what the nonlinear specification would be. Fourth, inspection of the data themselves may fail to suggest a clear alternative to the straight line model. (All too frequently, the scatterplot may look like nothing so much as a large chocolate chip cookie.) Below, we focus on establishing a linear relationship between variables. Nevertheless, we should always be alert to the possibility that a relationship is actually nonlinear.

Given that we want to relate Y to X with a straight line, the question arises as to which, out of all possible straight lines, we should choose. For the scatterplot of Figure 2a we have sketched in free-hand the line 1, defined by this prediction equation:

$$\hat{Y} = a_1 + b_1 X.$$

One observes that the line does not predict perfectly, for example, the vertical distance from Observation 1 to the line is three units. The *predic-*

tion error for this Observation 1, or any other observation, i, is calculated as follows:

$$\text{prediction error} = \text{observed} - \text{predicted} - Y_i - \hat{Y}_i.$$

Summing the prediction error for all the observations would yield a total prediction error (TPE), total prediction error = $\Sigma(Y_i - \hat{Y}_i) = (+3-3+4) = 4$.

Clearly, line 1 fits the data better than free-hand line 2 (see Figure 2b), represented by the equation,

$$\hat{Y} = a_2 + b_2X.$$

(TPE for line 2 = 18). However, there are a vast number of straight lines besides line 2 to which line 1 could be compared. Does line 1 reduce prediction error to the minimum, or is there some other line which could do better? Obviously, we cannot possibly evaluate all the free-hand straight lines that could be sketched on the scatterplot. Instead, we rely on the calculus, in order to discover the values of a and b which generate the line with the lowest prediction error.

Before presenting this solution, however, it is necessary to modify somewhat our notion of prediction error. Note that line 3 (see Figure 2c), indicated by the equation,

$$\hat{Y} = a_3 + b_3X,$$

provides a fit that is patently less good than line 1. Nevertheless, the TPE = 0 for line 3. This example reveals that TPE is an inadequate measure of error, because the positive errors tend to cancel out the negative errors (here, $-6-4 + 10 = 0$). One way to overcome this problem of opposite signs is to square each error. (We reject the use of the absolute value of the errors because it fails to account adequately for large errors and is computationally unwieldy.) Our goal, then, becomes one of selecting the straight line which *minimizes the sum of the squares of the errors* (SSE):

$$SSE = \Sigma(Y_i - \hat{Y}_i)^2.$$

Through the use of the calculus, it can be shown that this sum of squares is at a minimum, or "least," when the coefficients a and b are calculated as follows:

$$b = \frac{\Sigma(X_i - \bar{X})(Y_i - \bar{Y})}{\Sigma(X_i - \bar{X})^2}$$

$$a = \bar{Y} - b\bar{X}.$$

These values of a and b are our "least squares" estimates.

At this point it is appropriate to apply the least squares principle in a research example. Suppose we are studying income differences among local government employees in Riverside, a hypothetical medium-size midwestern city. Exploratory interviews suggest a relationship between income and education. Specifically, those employees with more formal training appear to receive better pay. In an attempt to verify whether this is so, we gather relevant data.

The Data

We do not have the time or money to interview all 306 employees on the city payroll. Therefore, we decide to interview a *simple random sample* of 32, selected from the personnel list which the city clerk kindly provided.[1] (The symbol for a sample is "n," so we can write n = 32.) The data obtained on the current annual income (labelled variable Y) and the number of years of formal education (labelled variable X) of each respondent is given in Table 2.

The Scatterplot

From simply reading the figures in Table 2, it is difficult to tell whether there is a relationship between education (X) and income (Y). However, the picture becomes clearer when the data are arranged in a *scatterplot*. In Figure 3, education scores are plotted along the X-axis, and income scores along the Y-axis. Every respondent is represented by a point, located where a perpendicular line from his or her X value intersects a perpendicular line from his or her Y value. For example, the dotted lines in Figure 3 fix the position of Respondent 3, who has an income of $6898 and six years of education.

Visual inspection of this scatterplot suggests the relationship is essentially linear, with more years of education leading to higher income. In equation form, the relationship appears as,

$$Y = a + bX + e,$$

TABLE 2
Data on Education and Income

Respondent	Education (in years) X	Income (in dollars) Y
1	4	$ 6281
2	4	10516
3	6	6898
4	6	8212
5	6	11744
6	8	8618
7	8	10011
8	8	12405
9	8	14664
10	10	7472
11	10	11598
12	10	15336
13	11	10186
14	12	9771
15	12	12444
16	12	14213
17	12	16908
18	12	18347
19	13	19546
20	14	12660
21	14	16326
22	15	12772
23	15	17218
24	16	12599
25	16	14852
26	16	19138
27	16	21779
28	17	16428
29	17	20018
30	18	16526
31	18	19414
32	20	18822

where Y = respondent's annual income (in dollars,), X = respondent's formal education (in years), a = intercept, b = slope, e = error.

Estimating this equation with least squares yields,

$$\hat{Y} = 5078 + 732 \, X,$$

which indicates the straight line that best fits this scatter of points (see Figure 4). This prediction equation is commonly referred to as a *bivariate regression equation*. (Further, we say Y has been "regressed on" X.)

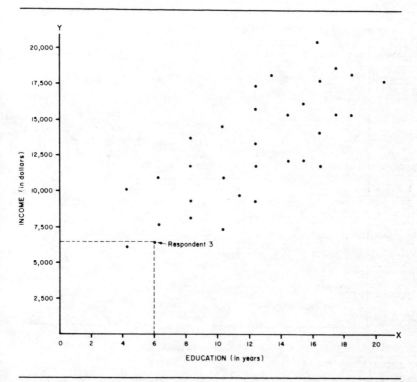

Figure 3: Scatterplot of Education and Income

The Slope

Interpretation of the estimates is uncomplicated. Let us first consider the estimate of the slope, b. *The slope estimate indicates the average change in Y associated with a unit change in X.* In our Riverside example, the slope estimate, 732, says that a one-year increase in an employee's amount of formal education is associated with an average annual income increase of $732. Put another way, we expect an employee with, say, 11 years of education to have an income that is $732 more than an employee having only 10 years of education. We can see how the slope dictates the change in Y for a unit change in X by studying Figure 4, which locates the expected values of Y, given X = 10, and X = 11, respectively.

Note that the slope tells us only the *average* change in Y that accompanies a unit change in X. The relationship between social science variables is inexact, that is, there is always error. For instance, we would not suppose that an additional year of education for any particular Riverside employee would be associated with an income rise of exactly $732. However, when we look at a large number of employees who have managed to

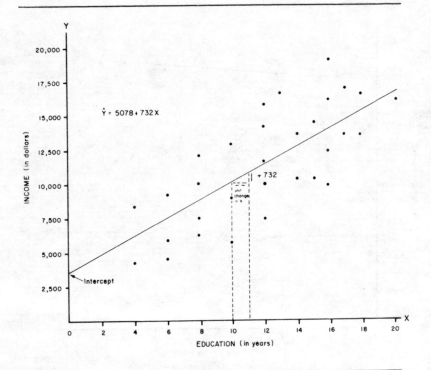

Figure 4: The Regression Line for the Income and Education Data

acquire this extra year of schooling, the average of their individual income gains would be about $732.

The slope estimate suggests the average change in Y *caused by* a unit change in X. Of course, this causal language may be inappropriate. The regression of Y on X might support your notion of the causal process, but it cannot establish it. To appreciate this critical point, realize that it would be a simple matter to apply least squares to the following regression equation,

$$X = a + bY + e,$$

where X = the *dependent* variable, Y = the *independent* variable. Obviously, such a computational exercise would not suddenly reverse the causal order of X and Y in the real world. The correct causal ordering of the variables is determined outside the estimation procedure. In practice, it is based on theoretical considerations, good judgment, and past research. With regard to our Riverside example, the actual causal relationship of these variables does seem to be reflected in our original model; that is, shifts in education appear likely to cause shifts in income; but, the view that changes in income cause changes in formal years of education is implausible, at least in this instance. Thus, it is only somewhat adventuresome to conclude that a one-year increase in formal education *causes* income to increase $732, on the average.

The Intercept

The intercept, a, is so called because it indicates the point where the regression line "intercepts" the Y-axis. It estimates the average value of Y when X equals zero. Thus, in our Riverside example, the intercept estimate suggests that the expected income for someone with no formal education would be $5078. This particular estimate highlights worthwhile cautions to observe when interpreting the intercept. First, one should be leery of making a prediction for Y based on an X value beyond the range of the data. In this example, the lowest level of educational attainment is four years; therefore, it is risky to extrapolate to the income of someone with zero years of education. Quite literally, we would be generalizing beyond the realm of our experience, and so may be way off the mark. If we are actually interested in those with no education, then we would do better to gather data on them.

A second problem may arise if the intercept has a negative value. Then, when X = 0, the predicted Y would necessarily equal the negative value. Often, however, in the real world it is impossible to have a score on Y that is below zero, for example, a Riverside employee could not receive a minus income. In such cases, the intercept is "nonsense," if taken literally. Its utility would be restricted to ensuring that a prediction "comes out right." It is a constant that must always be added on to the slope component, "bX," for Y to be properly estimated. Drawing on an analogy from the economics of the firm, the intercept represents a "fixed cost" that must be included along with the "varying costs" determined by other factors, in order to calculate "total cost."

Prediction

Knowing the intercept and the slope, we can predict Y for a given X value. For instance, if we encounter a Riverside city employee with 10 years of schooling, then we would predict his or her income would be $12,398, as the following calculations show:

$$\hat{Y} = 5078 + 732\,X$$
$$= 5078 + 732(10)$$
$$= 5078 + 7320$$
$$\hat{Y} = 12,398.$$

In our research, we might be primarily interested in prediction, rather than explanation. That is, we may not be directly concerned with identifying the variables that cause the dependent variable under study; instead, we may want to locate the variables that will allow us to make accurate guesses about the value of the dependent variable. For instance, in studying elections, we may simply want to predict winning candidates, not caring much about why they win. Of course, predictive models are not completely distinct from explanatory models. Commonly, a good explanatory model will predict fairly well. Similarly, an accurate predictive model is usually based on causal variables, or their surrogates. In developing a regression model, the research question dictates whether one emphasizes prediction or explanation. It is safe to conclude that, generally, social scientists stress explanation rather than prediction.

Assessing Explanatory Power: The R^2

We want to know how powerful an explanation (or prediction) our regression model provides. More technically, how well does the regression equation account for variations in the dependent variable? A preliminary judgment comes from visual inspection of the scatterplot. The closer the regression line to the points, the better the equation "fits" the data. While such "eyeballing" is an essential first step in determining the "goodness of fit" of a model, we obviously need a more formal measure, which the coefficient of determination (R^2) gives us.

We begin our discussion by considering the problem of predicting Y. If we only have observations on Y, then the best prediction for Y is always the estimated mean of Y. Obviously, guessing this average score for each case will result in many poor predictions. However, knowing the values of X, our predictive power can be improved, provided that X is related to Y. The question, then, is how much does this knowledge of X improve our prediction of Y?

In Figure 5 is a scatterplot, with a regression line fitted to the points. Consider prediction of a specific case, Y_1. Ignoring the X score, the best guess for the Y score would be the mean, $\bar{y}$. There is a good deal of error in this guess, as indicated by the deviation of the actual score from the mean, $Y_1 - \bar{Y}$. However, by utilizing our knowledge of the relationship of X to Y, we can better this prediction. For the particular value, X_1, the regression line predicts the dependent variable is equal to $\hat{Y}_1$, which is a clear improvement over the previous guess. Thus, the regression line has managed to account for some of the deviation of this observation from the mean; specifically, it "explains" the portion, $\hat{Y}_1 - \bar{Y}$. Nevertheless, our regression prediction is not perfect, but rather is off by the quantity, $Y_1 - \hat{Y}_1$; this deviation is left "unexplained" by the regression equation. In brief, the deviation of Y_1 from the mean can be grouped into the following components:

$$(Y_1 - \bar{Y}) \quad = \text{total deviation of } Y_1 \text{ from the mean, } \bar{Y}$$
$$(\hat{Y}_1 - \bar{Y}) \quad = \text{explained deviation of } Y_1 \text{ from } \bar{Y}$$
$$(Y_1 - \hat{Y}_1) \quad = \text{unexplained deviation of } Y_1 \text{ from } \bar{Y}.$$

We can calculate these deviations for each observation in our study. If we first square the deviations, then sum them, we obtain the complete components of variation for the dependent variable:

$$\Sigma(Y_i - \bar{Y})^2 = \text{total sum of squared deviations (TSS)}$$
$$\Sigma(\hat{Y}_i - \bar{Y})^2 = \text{regression (explained) sum of squared deviations (RSS)}$$
$$\Sigma(Y_i - \hat{Y}_i)^2 = \text{error (unexplained) sum of squared deviations (ESS).}$$

From this, we derive,

$$TSS = RSS + ESS.$$

The TSS indicates the total variation in the dependent variable that we would like to explain. This total variation can be divided into two parts: the part accounted for by the regression equation (RSS) and the part the regression equation cannot account for, ESS. (We recall that the least squares procedure guarantees that this error component is at minimum.) Clearly, the larger RSS relative to TSS, the better. This notion forms the basis of the R^2 measure:

$$R^2 = RSS / TSS.$$

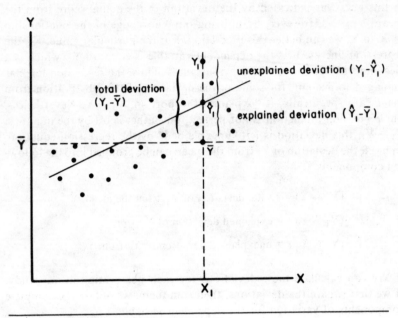

Figure 5: Components of Variation in Y

The coefficient of determination, R^2, indicates the explanatory power of the bivariate regression model. It records the proportion of variation in the dependent variable "explained" or "accounted for" by the independent variable. The possible values of the measure range from "+1" to "0." At the one extreme, when $R^2 = 1$, the independent variable completely accounts for variation in the dependent variable. All observations fall on the regression line, so knowing X enables the prediction of Y without error. Figure 6a provides an example where $R^2 = 1$. At the other extreme, when $R^2 = 0$, the independent variable accounts for no variation in the dependent variable. The knowledge of X is no help in predicting Y, for the two variables are totally independent of each other. Figure 6b gives an example where $R^2 = 0$ (note that the slope of the line also equals zero). Generally, R^2 falls between these two extremes. Then, the closer R^2 is to 1, the better the fit of the regression line to the points, and the more variation in Y is explained by X. In our Riverside example, $R^2 = .56$. Thus, we could say that education, the independent variable, accounts for an estimated 56% of the variation in income, the dependent variable.

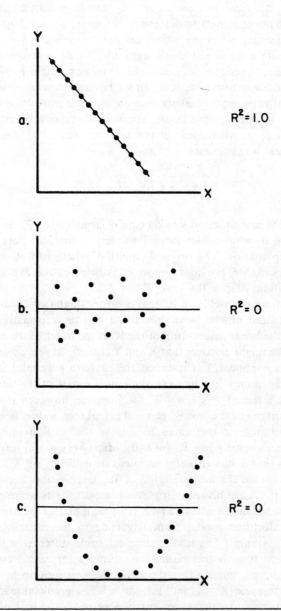

Figures 6a-c: Examples of the Extreme Values of the R^2

In regression analysis, we are virtually always pleased when the R^2 is high, because it indicates we are accounting for a large portion of the variation in the phenomenon under study. Further, a very high R^2 (say about .9) is essential if our predictions are to be accurate. (In practice, it is difficult to attain an R^2 of this magnitude. Thus, quantitative social scientists, at least outside economics, seldom make predictions.) However, a sizable R^2 does not necessarily mean we have a *causal* explanation for the dependent variable; instead, we may merely have provided a *statistical* explanation. In the Riverside case, suppose we regressed respondent's current income, Y, on income of the previous year, Y_{t-1}. Our revised equation would be as follows:

$$Y = a + bY_{t-1} + e.$$

The R^2 for this new equation would be quite large (above .9), but it would not really tell us what causes income to vary; rather, it offers merely a statistical explanation. The original equation, where education was the independent variable, provides a more convincing causal explanation of income variation, despite the lower R^2 of .56.

Even if estimation yields an R^2 that is rather small (say below .2), disappointment need not be inevitable, for it can be informative. It may suggest that the linear assumption of the R^2 is incorrect. If we turn to the scatterplot, we might discover that X and Y actually have a close relationship, but it is nonlinear. For instance, the curve (a parabola) formed by connecting the points in Figure 6c illustrates a perfect relationship between X and Y (i.e., $Y = X^2$), but $R^2 = 0$. Suppose, however, that we rule out nonlinearity. Then, a low R^2 can still reveal that X does help explain Y, but contributes a rather small amount to that explanation. Finally, of course, an extremely low R^2 (near 0), offers very useful information, for it implies that Y has virtually no linear dependency on X.

A final point on the interpretation of R^2 deserves mention. Suppose we estimate the *same* bivariate regression model for two samples from different populations, labelled 1 and 2. (For example, we wish to compare the income-education model from Riverside to the income-education model from Flatburg.) The R^2 for sample 1 could differ from the R^2 for sample 2, even though the parameter estimates for each were exactly the same. It simply implies that the structural relationship between the variables is the same ($a_1 = a_2$; $b_1 = b_2$), but it is less predictable in population 2. In other words, the same equation provides the best possible fit for both samples but, in the second instance, is less satisfactory as a total explanation of the dependent variable. Visually, this is clear. We can see, in comparing Figures 7a and 7b, that the points are clustered more tightly

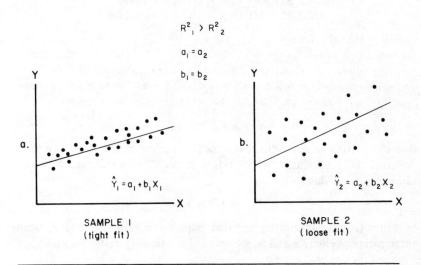

Figures 7a-b: Tight Fit vs. Loose Fit of a Regression Line

around the regression line of Figure 7a, indicating the model fits those data better. Thus, the independent variable, X, appears a more important determinant of Y in sample 1 than in sample 2.

R^2 Versus r

The relationship between the coefficient of determination. R^2, and the estimate of the correlation coefficient, r, is straightforward:

$$R^2 = r^2.$$

This equality suggests a possible problem with r, which is a commonly used measure of strength of association.[2] That is, r can inflate the importance of the relationship between X and Y. For instance, a correlation of .5 implies to the unwary reader that one-half of Y is being explained by X, since a perfect correlation is 1.0. Actually, though, we know that the r = .5 means that X explains only 25% of the variation in Y (because r^2 = .25), which leaves fully three-fourths of the variation in Y unaccounted for. (The r will equal the R^2 only at the extremes, when r = ± 1 or 0.) By relying on r rather than R^2, the impact of X on Y can be made to seem much greater than it is. Hence, to assess the strength of the relationship between the independent variable and the dependent variable, the R^2 is the preferred measure.

2. BIVARIATE REGRESSION:
ASSUMPTIONS AND INFERENCES

Recall that the foregoing regression results from the Riverside study are based on a *sample* of the city employees ($n = 32$). Since we wish to make accurate inferences about the actual *population* values of the intercept and slope parameters, this bivariate regression model should meet certain assumptions. For the population, the bivariate regression model is,

$$Y_i = \alpha + \beta X_i + \epsilon_i,$$

where the Greek letters indicate it is the population equation, and we have included the subscript, i, which refers to the ith observation. With the sample, we calculate

$$Y_i = a + bX_i + e_i.$$

In order to infer accurately the true population values, α and β, from these sample values, a and b, we make the following assumptions.

The Regression Assumptions

1. No specification error.

   a. The relationship between X_i and Y_i is linear.

   b. No relevant independent variables have been excluded.

   c. No irrelevant independent variables have been included.

2. No measurement error.

   a. The variables X_i and Y_i are accurately measured.

3. The following assumptions concern the error term, ϵ_i:

   a. Zero mean: $E(\epsilon_i) = 0$.

   i. For each observation, the *expected value* of the error term is zero. (We use the symbol E( ) for expected value which, for a random variable, is simply equal to its mean.)

   b. Homoskedasticity: $E(\epsilon_i^2) = 6^2$.

   i. The variance of the error term is constant for all values of X_i.

   c. No autocorrelation: $E(\epsilon_i \epsilon_j) = 0 \quad (i \neq j)$.

   i. The error terms are uncorrelated.

   d. The independent variable is uncorrelated with the error term: $E(\epsilon_i X_i) = 0$.

   e. Normality.

   i. The error term, ϵ_i, is normally distributed.

When assumptions 1 to 3d are met, desirable estimators of the population parameters, α and β, will be obtained; technically, they will be the "best linear unbiased estimates," BLUE. (An unbiased estimator correctly estimates the population parameter, on the average, i.e., $E(b) = \beta$. For instance, if we repeatedly draw samples from the population, each time recalculating b, we would expect the average of all these b's to equal β.) If the normality assumption (3e) also holds, they will be the "best unbiased estimates," and we can carry out significance tests, in order to determine how likely it is that the population parameter values differ from zero. Below, we consider each assumption in more detail.

The first assumption, absence of specification error, is critical. In sum, it asserts that the theoretical model embodied in the equation is correct. That is, the functional form of the relationship is actually a straight line, and no variables have been improperly excluded or included as "causes." Let us examine the Riverside example for specification error. Visual inspection of the shape of the scatterplot (see Figure 4), along with the $R^2 = .56$, indicates that the relationship is essentially linear. However, it seems likely that relevant variables have been excluded, for factors besides education undoubtedly influence income. These other variables should be identified and brought into the equation, both to provide a more complete explanation and to assess the impact of education after additional forces are taken into account. (We take up this task in the next chapter.) The final aspect of specification error, inclusion of an irrelevant variable, argues that education might not really be associated with income. To evaluate this possibility, we will perform a test for statistical significance.

The need for the second assumption, no measurement error, is self-evident. If our measures are inaccurate, then our estimates are likely to be inaccurate. For instance, with the Riverside case, suppose that in the measurement of the education variable, the respondents tended to report the number of years of schooling they would *like* to have had, rather than the number of years of schooling they *actually* had. If we were to use such a variable to indicate actual years of schooling, it would contain error, and the resulting regression coefficient would not accurately reflect the impact of actual education on income. When the analyst cannot safely rule out the possibility of measurement error, then the magnitude of the estimation problem depends on the nature and location of the error. If only the dependent variable is measured with error, then the least squares estimates may remain unbiased, provided the error is "random.' However, if the independent variable is measured with any error, then the least squares estimates will be biased. In this circumstance, all solutions are problematic. The most oft-cited approach is *instrumental vari-*

ables estimation, but it cannot promise the restoration of unbiased parameter estimates.

The third set of assumptions involve the error term. The initial one, a zero mean, is of little concern because, regardless, the least squares estimate of the slope is unchanged. It is true that, if this assumption is not met, the intercept estimate will be biased. Nevertheless, since the intercept estimate is often of secondary interest in social science research, this potential source of bias is rather unimportant.

Violating the assumption of homoskedasticity is more serious. While the least squares estimates continue to be unbiased, the significance tests and confidence intervals would be wrong. Let us examine Figure 4 from the Riverside study. Homoskedasticity would appear to be present, because the variance in prediction errors is more or less constant across the values of X; that is, the points snuggle in a band of equal width above and below the regression line. If, instead, the points fanned out from the regression line as the value of X increased, the assumption would not hold, and a condition of *heteroskedasticity* would prevail. The recommended solution for this condition is a *weighted least squares* procedure. (Diagnosis of heteroskedasticity is discussed further when the analysis of residuals is considered.)

The assumption of no autocorrelation means that the error corresponding to an observation is not correlated with any of the errors for the other observations. When autocorrelation is present, the least squares parameter estimates are still unbiased; however, the significance tests and confidence intervals are invalid. Commonly, significance tests will be much more likely to indicate that a coefficient is statistically significant, when in fact it is not. Autocorrelation more frequently appears with *time-series* variables (repeated observations on the same unit through time) than with *cross-sectional* variables (unique observations on different units at the same point in time, as with our Riverside study). With time-series data, the no autocorrelation assumption requires that error for an observation at an earlier time is not related to errors for observations at a later time. If we conceive of the error term in the equation as, in part, a summary of those explanatory variables that have been left out of the regression model, then no autocorrelation implies that those forces influencing Y in, say, year 1, are independent of those forces influencing Y in year 2.[3] This assumption, it should be obvious, is often untenable. (The special problems of time-series analysis have generated an extensive literature; for a good introduction, see Ostrom, 1978.)

The next assumption, that the independent variable is uncorrelated with the error term, can be difficult to meet in nonexperimental research. Typically, we cannot freely set the values of X like an experimenter would, but rather must merely observe values of X as they present themselves in society. If this observed X variable is related to the error term, then the least squares parameter estimates will be biased. The simplest way to test for this violation is to evaluate the error term as a collection of excluded explanatory variables, each of which might be correlated with X. Thus, in the Riverside case, the error term would include the determinants of income other than education, such as sex of the respondent. If the explanatory variable of education is correlated with the explanatory variable of sex, but this latter variable is excluded from the equation, then the slope estimate for the education variable in the bivariate regression will be biased. This b will be too large, because the education variable is allowed to account for some income variation that is actually the product of sex differences. The obvious remedy, which we come to employ, is the incorporation of the missing explanatory variables into the model. (If, for some reason, an explanatory variable cannot be so incorporated, then we must trust the assumption that, as part of the error term, it is uncorrelated with the independent variable actually in the model.)

The last assumption is that the error term is normally distributed. Since the distributions of Y_i and ϵ_i are the same (only their means are different), our discussion will be facilitated by simply considering the distribution of Y_i. The frequency distribution of a variable that conforms to a normal curve has a symmetric bell-shape, with 95% of the observations falling within two standard deviations, plus or minus, of the mean. With regard to the Riverside example, the unique observations on the income variable (Y_i) could be graphed onto a frequency polygon to allow a visual inspection for normality. Or, for a quick preliminary check, we could count the number of observations above and below the mean, expecting about half in either direction. (In fact, there are 16 incomes above and 16 incomes below the mean of \$13,866, which suggests a normal distribution.) A more formal measure, which takes into account all the information in the frequency distribution, is the *skewness* statistic, based on the following formula:

$$\text{skewness} = \frac{\Sigma \left(\dfrac{y_i - \overline{y}}{s_y} \right)^3}{n}.$$

If the distribution is normal, then skewness = 0. For our income variable, skewness measures only −.02, indicating that the distribution is virtually normal.

There is some disagreement in the statistical literature over how serious the violations of the regressions assumptions actually are. At one extreme, researchers argue that regression analysis is "robust," that is, the parameter estimates are not meaningfully influenced by violations of the assumptions. This "robust" perspective on regression is employed in Kerlinger and Pedhazar (1973). At the other extreme, some feel that violations of the assumptions can render the regression results almost useless. Bibby's (1977) work provides an example of this "fragile" view of regression analysis. Clearly, some of the assumptions are more robust than others. The normality assumption, for instance, can be ignored when the sample size is large enough, for then the central-limit theorem can be invoked. (The central-limit theorem indicates that the distribution of a sum of independent variables, which we can conceive of the error term as representing, approaches normality as sample size increases, irrespective of the nature of the distributions in the population.) By way of contrast, the presence of specification error, such as the exclusion of a relevant variable, creates rather serious estimation problems which can be relieved only by introduction of the omitted variable into the model. Those who wish to gain a fuller understanding of this controversy over assumptions should consult, in addition to the efforts just cited, the excellent paper by Bohrnstedt and Carter (1971). More advanced treatments of the regression assumptions are available in econometrics texts; listing them in order of increasing difficulty, I would recommend Kelejian and Oates (1974), Pindyck and Rubinfeld (1976), and Kmenta (1971).

Confidence Intervals and Significance Tests

Because social science data invariably consist of samples, we worry whether our regression coefficients actually have values of zero in the population. Specifically, is the slope (or the intercept) estimate significantly different from zero? (Of course, we could test whether the parameter estimate was significantly different from some number other than zero; however, we generally do not know enough to propose such a specific value.) Formally, we face two basic hypotheses: the null and an alternative. The *null hypothesis* states that X is not associated with Y; therefore, the slope, β, is zero in the population. An *alternative hypothesis* states that

X is associated with Y; therefore, the slope is *not* zero in the population. In summary, we have

$$H_0: \beta = 0 \text{ (null hypothesis)}$$

$$H_1: \beta \neq 0 \text{ (alternative hypothesis).}$$

To test these hypotheses, an interval can be constructed around the slope estimate, b. The most widely used is a two-tailed, 95% *confidence interval*:

$$(b \pm t_{n-2;.975} s_b).$$

If the value of zero does *not* fall within this interval, we reject the null hypothesis and accept the alternative hypothesis, with 95% confidence. Put another way, we could conclude that the slope estimate, b, is significantly different from zero, at the .05 level. (The level of *statistical significance* associated with a particular confidence interval can be determined simply by subtracting the confidence level from unity, for example, $1 - .95 = .05$.)

In order to apply this confidence interval, we must understand the terms of the formula. These are easy enough. The term s_b is an estimate of the standard deviation of the slope estimate, b, and is commonly referred to as the *standard error*. It is a useful measure of the dispersion of our slope estimate. The formula for this standard error is,

$$s_b = \sqrt{\frac{\Sigma(Y - \hat{Y})^2/(n-2)}{\Sigma(X - \bar{X})^2}}.$$

Statistical computing packages such as SPSS routinely print out the standard errors when estimating a regression equation.

Because s_b is an estimate (we seldom actually know the standard deviation of the slope estimate), it is technically incorrect to use the normal curve to construct a confidence interval for β. However, we can utilize the t distribution with (n−2) degrees of freedom. (The t distribution is quite similar to the normal distribution, especially as n becomes large, say greater than 30.) Almost every statistics text provides a table for the t distribution.

The last component in the confidence interval formula is the subscript, ".975." This merely indicates that we are employing a 95% confidence interval, but with two-tails. A two-tailed test means that the hypothesis about the affect of X on Y is nondirectional; for example, the above alternative hypothesis, H_1, is sustained if b is either significantly negative or significantly positive.

Suppose we now construct a two-tailed, 95% confidence interval around the regression coefficients in our Riverside study. We have,

$$\hat{Y} = 5078 + 732X$$
$$\qquad (1498) \quad (118)$$

where the figures in parentheses are the standard errors of the parameter estimates. Given the sample size is 32,

$$t_{n-2;.975} = t_{32-2;.975} = t_{30;.975} = 2.04,$$

according to the t table. Therefore, the two-tailed, 95% confidence interval for β is

$$(b \pm t_{n-2;.975}s_b) = 732 \pm 2.04 \,(118) = (732 \pm 241).$$

The probability is .95 that the value of the population slope, β, is between $491 and $973. Since the value of zero does not fall within the interval, we reject the null hypothesis. We conclude that the slope estimate, b, is significantly different from zero, at the .05 level.

In the same fashion, we can construct a confidence interval for the intercept, α. Continuing the Riverside example,

$$(a \pm t_{n-2;.975}s_a) = 5078 \pm 2.04\,(1498) = (5078 \pm 3056).$$

Clearly, the two-tailed, 95% confidence band for the intercept does not contain zero. We reject the null hypothesis and declare that the intercept estimate, a, is statistically significant at the .05 level. Graphically, this means we reject the possibility that the regression line cuts the origin.

Besides providing significance tests, confidence intervals also allow us to present our parameter estimates as a range. In a bivariate regression equation, b is a *point estimate*; that is, it is a specific value. The confidence band, in contrast, gives us an *interval estimate*, indicating that the slope in the population, β, lies within a range of values. We may well choose to stress the interval estimate over the point estimate. For example, in our Riverside study the point estimate of β is $732. This is our best guess, but in reporting the results we might prefer to say merely that a year increase in education is associated with an increase of "more or less $732" a year in income. Estimating a confidence interval permits us to formalize this caution; we could assert, with 95% certainty, that a one-year increase in education is associated with an income increase of from $491 to $973.

In the Riverside investigation, we have rejected, with 95% confidence, the null hypothesis of no relationship between income and education. Still, we know that there is a 5% chance we are wrong. If, in fact, the null hypothesis is correct, but we reject it, we commit a *Type I error*. In an effort to avoid Type I error, we could employ a 99% confidence interval, which broadens the acceptance region for the null hypothesis. The formula for a two-tailed, 99% confidence interval for β is as follows:

$$(b \pm t_{n-2;.995}s_b).$$

Applying the formula to the Riverside example,

$$732 \pm 2.75\,(118) = (732 \pm 324).$$

These results provide some evidence that we have not committed a Type I error. This broader confidence interval does not contain the value of zero. We continue to reject the null hypothesis, but with greater confidence. Further, we can say that the slope estimate, b, is statistically significant at the .01 level. (This effort to prevent Type I error involves a trade-off, for the risk of *Type II error*, accepting the null hypothesis when it is false, is inevitably increased. Type II error is discussed below.)

The One-Tailed Test

Thus far, we have concentrated on a two-tailed test of the form,

$$H_0: \beta = 0$$
$$H_1: \beta \neq 0.$$

Occasionally, though, our acquaintance with the phenomena under study suggests the sign of the slope. In such a circumstance, a one-tailed test might be more reasonable. Taking the Riverside case, we would not expect the sign of the slope to be negative, for that would mean additional education actually decreased income. Therefore, a more realistic set of hypotheses here might be,

$$H_0: \beta = 0$$
$$H_1: \beta > 0.$$

Applying a one-tailed, 95% confidence interval yields,

$$\beta > (b - t_{n-2;.95}s_b) = 732 - 1.70\,(118) = (732 - 201) = 531.$$

The lower boundary of the interval is above zero. Therefore, we reject the null hypothesis and conclude the slope is positive, with 95% confidence.

Once the level of confidence is fixed, it is "easier" to find statistical significance with a one-tailed test, as opposed to a two-tailed test. (The two-tailed confidence interval is more likely to capture zero. For instance, the lower bounds in the Riverside case for the two-tailed and one-tailed tests, respectively, are \$491 and \$531.) This makes intuitive sense, for it takes into account the researcher's prior knowledge, which may rule out one of the tails from consideration.

Significance Testing: A Rule of Thumb

Recall the formula for the two-tailed, 95% confidence interval for β:

$$(b \pm t_{n-2;.975}s_b).$$

If this confidence interval does not contain zero, we conclude that b is significant at the .05 level. We see that this confidence interval will not contain zero if, when b is positive,

$$(b - t_{n-2;.975}s_b) > 0,$$

or, when b is negative,

$$(b + t_{n-2;.975}s_b) < 0.$$

These requirements may be restated as,

$$b/s_b > t_{n-2;.975}, \text{ when b is positive,}$$

or,

$$b/s_b < t_{n-2;.975}, \text{ when b is negative.}$$

In brief, these requirements can be written,

$$|b/s_b| > t_{n-2;.975},$$

which says that when the absolute value of the parameter estimate, b, divided by its standard error, s_b, surpasses the t distribution value, $t_{n-2;.975}$, we reject the null hypothesis. Thus, a significance test at the .05 level, two-tailed, can be administered by examining this ratio. The test is simplified further when one observes that, for almost any sample size, the value in the t distribution approximates 2. For example, if the sample size is only 20, then $t_{20-2;.975} = t_{18;.975} = 2.10$. In contrast, if the sample is of infinite size, $t_{\infty;.975} = 1.96$. This narrow range of values given by the t distribution leads to the following rule of thumb. If,

$$|b/s_b| > 2,$$

then the parameter estimate, b, is significant at the .05 level, two-tailed.

This *t ratio*, as it is called, is routinely printed in the regression component of several computer data analysis programs. Otherwise, it is easily calculated by dividing b by s_b. The t ratio provides an efficient means of significance testing, and researchers frequently employ it. Of course, whenever more precision is wanted, the t table can always be consulted. Below is the bivariate regression model from our Riverside example, with the t ratios appearing in parentheses under the parameter estimates:

$$\hat{Y} = 5078 + 732X$$
$$(3.39) \quad (6.23).$$

A quick glance at the t ratios reveals they exceed 2; we immediately conclude that both a and b are statistically significant at the .05 level.

Reasons Why a Parameter Estimate May Not Be Significant

There are many reasons why a parameter estimate may be found not significant. Let us assume, to narrow the field somewhat, that our data compose a probability sample and that the variables are correctly measured. Then, if b turns out not to be significant, the most obvious reason is that X is not a cause of Y. However, suppose we doubt this straightforward conclusion. The following is a partial list of reasons why we might fail to uncover statistical significance, even though X is related to Y in fact:

(1) inadequate sample size
(2) Type II error
(3) specification error
(4) restricted variance in X.

Below, these four possibilities are evaluated in order. (A fifth possibility is high multicollinearity, which we will consider in our discussion of multiple regression.)

As sample size increases, a given coefficient is more likely to be found significant. For insance, the b value in the bivariate regression of the Riverside example would not be significant (.05) if based on only five cases, but is significant with n = 32. This suggests it may be worthwhile for a researcher to gather more observations, for it will be easier to detect a relationship between X and Y in the population, if one is present. In fact, with a very large sample, statistical significance can be uncovered even if b is substantively quite small. (For very large samples, such as election surveys of 1000 or more, significance may actually be "too easy" to find, since tiny coefficients can be statistically significant. In this situation, the analyst might prefer to rely primarily on a substantive judgment of the importance of the coefficient.)

Let us suppose that sample size is fixed, and turn to the problem of choosing a significance level, as it relates to Type II error. In principle, we could set the significance test at any level between 0 and 1. In practice, however, most social scientists employ the .05 or .01 levels. To avoid the charges of arbitrariness or bias, we normally select one of these conventional standards before analysis begins. For instance, suppose prior to our investigation we decide to employ the .01 significance level. Upon analysis, we find b is not significant at this .01 level. But, we observe that it is significant at the less demanding level of .05. We might be loath to accept the null hypothesis as dictated by the .01 test, especially since theory and prior research indicate that X does influence Y. Technically, we worry that we are committing a Type II error, accepting the null when it is false. In the end, we may prefer to accept the results of the .05 test. (In this particular case, given the strength of theory and previous research, perhaps we should have initially set the significance test at the less demanding .05 level.)

Aside from Type II error, b may not appear significant because the equation misspecifies the relationship between X and Y. Perhaps the relationship follows a curve, rather than a straight line, as assumed by the regression model. First, this curvilinearity should be detectable in the scatterplot. To establish the statistical significance of the relationship in the face of this curvilinearity, regression analysis might still be applied, but the variables would have to be properly transformed. (We pursue an example of such a transformation of the end of this chapter.)

Finally, a parameter estimate may not be found significant because the variance in X is restricted. Look again at the formula for the standard error of b, s_b.

$$s_b = \sqrt{\frac{\Sigma(Y - \hat{Y})^2/(n-2)}{\Sigma(X - \bar{X})^2}}.$$

We can see that as the dispersion of X about its mean decreases, the value of the denominator decreases, thereby increasing the standard error of b. Other things being equal, a larger standard error makes statistical significance more difficult to achieve, as the t ratio formula makes clear. The implication is that b may not be significant simply because there is too little variation in X. (The degree of variation in X is easily checked by evaluating its standard deviation.) In such a circumstance, the researcher may seek to gather more extreme observations on X, before making any firm conclusions about whether it is significantly related to Y.

The Prediction Error for Y

In regression analysis, the difference between the observed and the estimated value of the dependent variable, $Y_i - \hat{Y}_i$, equals the prediction error for that case. The variation of all these prediction errors around the regression line can be estimated as follows:

$$s_e = \sqrt{\frac{\Sigma(Y_i - \hat{Y}_i)^2}{n-2}}.$$

This s_e is called the *standard error of estimate of Y*; that is, the estimated standard deviation of the actual Y from the predicted Y. Hence, the standard error of estimate of Y provides a sort of average error in predicting Y. Further, it can be used to construct a confidence interval for Y, at a given X value. Utilizing the knowledge that the value given by the t distribution approximates 2 for a sample of almost any size, we produce the following 95% confidence interval for Y:

$$(\hat{Y} \pm 2s_e).$$

Let us take an example. In the Riverside study, we would predict someone with 10 years of education had an income of

$$\hat{Y} = 5078 + 732(10) = 12,398.$$

How accurate is this prediction? For X = 10, we have this 95% confidence
interval (s_e = 2855):

$$12,398 \pm 2(2855) = (12,398 \pm 5710).$$

According to this confidence interval, there is a .95 probability that a
city employee with 10 years of education has an annual income between
$6688 and $18,108. This is not a very narrow range of values. (The high
extreme is almost three times the low extreme.) We conclude that our
bivariate regression model cannot predict Y very accurately, for a specific
value of X. Such a result is not too surprising. Recall that, according to
the R^2 = .56, the model explains just over one-half the variation in Y. Our
R^2 would need to be much greater, in order to reduce our prediction error
substantially.

A last point merits mention. The above confidence interval, which
utilizes s_e, provides a kind of "average" confidence interval. In reality, as
the value of X departs from the mean, the actual confidence interval
around Y tends to get larger. Thus, at more extreme values of X, the
above confidence interval will be somewhat narrower than it should be.
The formula for constructing this more precise confidence interval is
readily available (see Kelejian and Oates, 1974, pp. 111-116).

Analysis of Residuals

The prediction errors from a regression model, $Y_i - \hat{Y}_i$, are also called
residuals. Analysis of these residuals can help us detect the violation of
certain of the regression assumptions. In a visual inspection of the residuals, we hope to observe a healthy pattern similar to that in Figure 8a;
that is, the points appear scattered randomly about in a steady band of
equal width above and below the regression line. Unfortunately, however,
we might discover a more problematic pattern resembling one of those
in Figures 8b to 8e. Below, we consider each of these troublesome patterns, in turn.

We begin with the most easily detectable problem, that of *outliers*.
In Figure 8b, there are two observations with extremely large residuals,
placing them quite far from the regression line. At least with regard to
these observations, the linear model provides a very poor fit. By looking

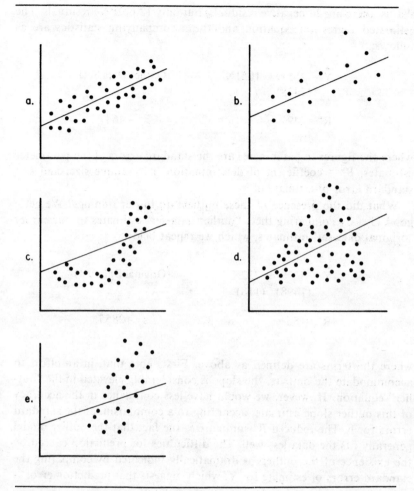

Figures 8a-e: Some Possible Patterns for Residuals

at a concrete example, we can explore consequences of outliers in more detail. In our Riverside study, suppose we had been careless in coding the data and recorded the incomes of Respondents 29 and 30 as $30,018 and $36,526, respectively (instead of the correct values, $20,018 and $16,526). The scatterplot, adjusted to include these erroneous values, would now look like Figure 9. By fitting a regression line to this revised plot, we see that Respondents 29 and 30 have become outliers, with residuals of 10,112 and 15,599, respectively. Further, examining the residuals generally, we note that they are out of balance around the line, that is, there are 20 negative residuals, but only 12 positive residuals. The estimated regression equation and the accompanying statistics are as follows:

$$\hat{Y} = 2557 + 1021X \qquad \text{(Outlier Data-Set)}$$
$$(2438) \quad (191)$$

$$R^2 = .49 \qquad n = 32 \qquad s_e = 4647,$$

where the figures in parentheses are the standard errors of the parameter estimates; R^2 = coefficient of determination, n = sample size; and s_e = standard error of estimate of Y.

What did the presence of these outliers do to our findings? We get a good idea by comparing these "outlier data-set" estimates to our earlier "original data-set" estimates, which we repeat below:

$$\hat{Y} = 5078 + 732X \qquad \text{(Original Data-Set)}$$
$$(1498) \quad (118)$$

$$R^2 = .56 \qquad n = 32 \qquad s_e = 2855,$$

where the terms are defined as above. First, note that, in an effort to accommodate the outliers, the slope is considerably elevated in the "outlier" equation. However, we would have less confidence in the accuracy of this outlier slope estimate, according to a comparison of the standard errors for b. The reduced R^2 summarizes the fact that the outlier model generally fits the data less well. The difficulties for prediction caused by the existence of the outliers is dramatically indicated by comparing the standard errors of estimate for Y, which suggests that prediction error is over 1.5 times as great under the outlier equation.

These several statistics show that the presence of outliers clearly weakens our explanation of Y. How can we adjust for outliers, in general? (We refer, of course, to actual outliers, not outliers that could be corrected by more careful coding, as in our pedagogic example.) There are at least four possibilities:

(1) Exclude the outlying observations.
(2) Report two equations, one with the outliers included and one without.
(3) Transform the variable.
(4) Gather more observations.

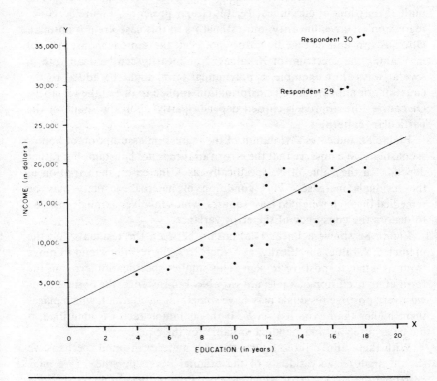

Figure 9: The Fitted Regression Line in the Presence of Outliers

There are pros and cons attached to each of these possibilities. Adjustment 1 simply eliminates the problem by eliminating the outliers. The principle drawbacks are the reduction in sample size and the loss of information it entails. Adjustment 2 preserves the information that would be lost in Adjustment 1; however, it may be cumbersome to have to consider two empirically different versions of ostensibly the same model. Adjustment 3 uses only one equation, maintains the sample size, and can pull the outliers closer to the regression line. However, the results may be robbed of the straightforward interpretation possible when the variable was measured in the original units. Adjustment 4 may reveal that the outliers are not atypical cases, but in fact fit into a more general, perhaps nonlinear, pattern. An obvious limitation is that usually, in nonexperimental social science research, it is impossible to gather more observations. None of these adjustments is appropriate for every situation. Rather, in deciding on how to handle an outlier problem, we must consider our research question and the appearance of the particular scatterplot.

Figures 8c to 8e represent more "abnormal" residual plots. While outliers may hint at curvilinearity, it is clearly present in Figure 8c. Since regression assumes linearity, our estimators in this case are not optimal. Obviously, a unit change in X does not elicit the same response (i.e., b) in Y along the spectrum of X values. Nonlinearity can be dealt with in several ways. For example, a polynomial term might be added to the equation, or a logarithmic transformation applied to one of the variables. Of course, the approach chosen depends partly upon the shape of the particular scatterplot.

Figure 8d indicates a violation of the regression assumption of homoskedasticity. We observe that the error variance is not constant, but rather depends on the value of X; specifically, as X increases, the variation of the residuals increases. This condition of heteroskedasticity may be remedied through weighted least squares, which involves a transformation to restore the constancy of the error variance.

Figure 8e shows a linear relationship between the residuals and the predicted Y value; specifically, as Y increases, the residuals tend to move from negative to positive in sign. This implies specification error in the form of an exclusion of a relevant variable. For instance, the observations with very positive residuals may have something in common which places them higher than expected on Y. If this common factor is identified, it indicates a second independent variable for the equation.

With these above three figures (8c, 8d, and 8e) in mind, perhaps we should analyze the residuals of the original Riverside study. (We have corrected the coding error which produced the outliers.) Of course, these residuals could be examined simply by looking at the scatter around the regression line, as we have done thus far. Sometimes, however, we want to highlight them in a special plot. Figure 10 shows such a plot, where the residual values are indicated on the vertical axis, the predicted Y values are indicated on the horizontal axis. This residual plot fails to suggest any of the patterns in Figures 8c to 8e. The residuals neither follow a curve, nor do they take the shape of the heteroskedastic "fan." Also, if there is specification error, it cannot be detected through analysis of these residuals. In sum, the pattern of residuals in Figure 10 appears free of abnormalities, forming a straight, broad band which the horizontal line cuts in half. This visual impression receives quantitative confirmation. A simple sign count reveals a virtually even balance around the line (17 negative residuals, 15 positive residuals). Further, all the residuals are scattered within a band that extends from the line plus or minus two standard errors of estimate of Y.

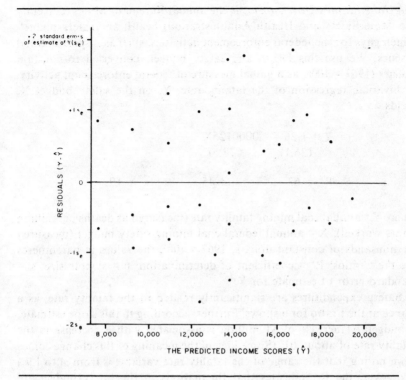

Figure 10: A Plot of Residuals

The Effect of Safety Enforcement on Coal Mining
Fatalities: A Bivariate Regression Example

It is time to apply what we have learned to some data from the real world. A current public policy controversy concerns whether the federal government can regulate safety in the workplace. Before the 1970 passage of the Occupational Safety and Health Act, federal government involvement in occupational safety was limited to coal mining. A study of this intervention, which extends over 35 years, may shed light on the act's prospects for success. Our specific research question is, "Has federal safety enforcement reduced the rate of fatalities in the coal mines?" From various issues of the *Minerals Yearbook*, annual observations, 1932-1976, can be gathered on the U.S. coal mining fatality rate (measured as number of deaths per million hours worked). Also available, from *The Budget of the United States Government*, is the annual Bureau of Mines (currently the Mine Safety and Health Administration) health and safety budget, which pays for the federal enforcement activities, such as inspections and rescues. We use this health and safety budget, converted to constant dollars (1967 = 100), as a global measure of federal enforcement activity. A bivariate regression of the fatality rate, Y, on the safety budget X, yields

$$\hat{Y} = 1.26 - .0000125X$$
$$(36.1) \qquad (-8.5)$$

$$R^2 = .63 \qquad n = 45 \qquad s_e = .19,$$

where Y = annual coal mining fatality rate (measured as deaths per million hours worked), X = annual federal coal mining safety budget (measured in thousands of constant dollars, 1967 = 100); the values in parentheses are the t ratios; R^2 = coefficient of determination; n = sample size; s_e = standard error of estimate for Y.

Safety expenditures are significantly related to the fatality rate, as a glance at the t ratio for b shows. Further, according to this slope estimate, a budgetary increase of $1 million is associated with a decrease in the fatality rate of about .01. (Some idea of the meaning of this change comes from noting that the range of the fatality rate variable is from .4 to 1.7.) Moreover, the R^2 indicates that fluctuations in the safety budget are responsible for over one-half the variation in the fatality rate. In sum, federal safety enforcement, as measured by expenditures for that purpose, seems an important influence on the coal mining fatality rate.

These estimates, although they appear pleasing, should not be accepted too readily, for we failed to look at the scatterplot. Upon inspection we discover that the linearity assumed by our regression equation is incorrect. Rather, the relationship between X and Y tends to follow a curve, as sketched in Figure 11. Fortunately, we are often able to transform the variables so as to make a relationship linear. The shape of this curve strongly suggests a logarithmic transformation is the most appropriate. Specifically, a logarithmic transformation of X will tend to "straighten out" the scatter, thus rendering the data more compatible with the linear regression assumption. Further, this transformation incorporates the knowledge gleaned from Figure 11, which is that, contrary to the interpretation from the above slope estimate, each additional dollar spent decreases the fatality rate *less and less*. (For an excellent discussion of logarithmic transformations, see Tufte, 1974, pp. 108-131.)

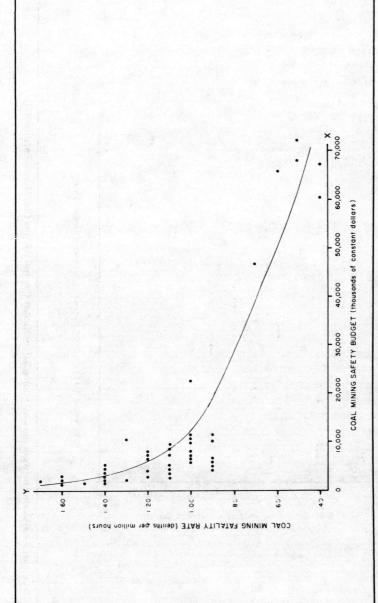

Figure 11: Curvilinear Relationship between the Coal Mining Safety Budget and the Coal Mining Fatality Rate

45

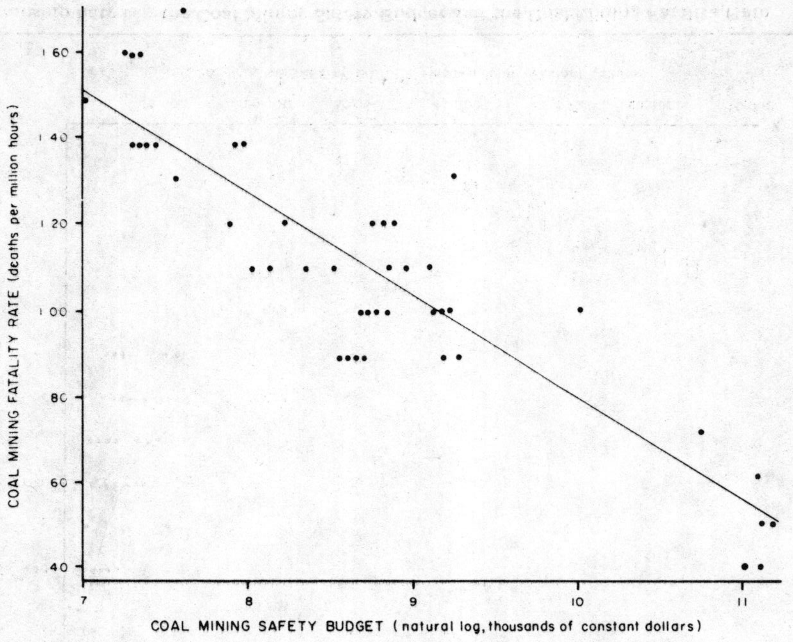

Figure 12: The Linear Relationship between the Coal Mining Safety Budget (logged) and the Coal Mining Fatality Rate

46

Figure 12 shows the new scatterplot, after X has undergone a "natural" logarithmic transformation, $\ell n\ X$. Reestimating the equation, but with this transformed X, yields,

$$\hat{Y} = 3.25 - \underset{(-13.6)}{.247\ \ell n\ X}$$
$$\phantom{\hat{Y} = }(20.3)$$

$$R^2 = .81 \qquad n = 45 \qquad s_e = .14,$$

where the terms are defined as above.

Our explanation of the fatality rate is considerably improved. This equation accounts for over two-thirds of the variation in Y, as the R^2 reveals. Further, the increment in R^2 from the earlier equation is large (.81 − .63 = .18), demonstrating that the curvilinearity in the relationship of safety expenditures to the fatality rate is substantial. Incorporating this curvilinearity into our model markedly enhances the predictive power of the model. In the earlier equation, when Y is predicted for a given budget, the average error is .19. This standard error of estimate for Y is reduced to .14 in our revised model. By careful examination of the original scatterplot and application of the appropriate transformation, we noticeably bettered what, at first blush, appeared to be an adequate accounting of the association between coal mining fatalities and federal safety expenditures. Of course, although safety expenditures represent an important determinant of the fatality rate, it is not the only one, as we will discover in the next chapter.

3. MULTIPLE REGRESSION

With multiple regression, we can incorporate more than one independent variable into an equation. This is useful in two ways. First, it almost inevitably offers a fuller explanation of the dependent variable, since few phenomena are products of a single cause. Second, the effect of a particular independent variable is made more certain, for the possibility of distorting influences from the other independent variables is removed. The procedure is a straightforward extension of bivariate regression. Parameter estimation and interpretation follow the same principles. Likewise, the significance test and the R^2 are parallel. Further, the bivariate regression assumptions necessary for BLUE are carried over to the multivariate case. The technique of multiple regression has great range, and its mastery will enable the researcher to analyze virtually any set of quantitative data.

The General Equation

In the general multiple regression equation, the dependent variable is seen as a linear function of more than one independent variable,

$$Y = a_0 + b_1X_1 + b_2X_2 + b_3X_3 + \ldots + b_kX_k + e,$$

where the subscript identifies the independent variables. The elementary three-variable case, which we shall be using below, is written,

$$Y = a_0 + b_1X_1 + b_2X_2 + e,$$

and suggests that Y is determined by X_1 and X_2, plus an error term.

To estimate the parameters we again employ the least squares principle, minimizing the sum of the squares of the prediction errors (SSE):

$$SSE = \Sigma(Y - \hat{Y})^2.$$

For the three-variable model, this least squares equation is,

$$\hat{Y} = a_0 + b_1X_1 + b_2X_2.$$

The least squares combination of values for the coefficients (a_0, b_1, b_2) yields less prediction error than other possible combinations of values. Hence, the least squares equation fits the set of observations better than any other linear equation. However, it can no longer be represented graphically with a simple straight line fitted to a two-dimensional scatterplot. Rather, we must imagine fitting a *plane* to a three-dimensional scatter of points. The location of this plane, of course, is dictated by the values of a_0, b_1, and b_2, which are given by the calculus. For most of us, it is impossible to visualize the fitting of equations with more than three variables. Indeed, for the general case, with k independent variables, it requires conceiving of adjusting a k-dimensional hyperplane to a (k + 1)-dimensional scatter.

For purposes of illustration, let us look at a simple three-variable model from our Riverside study. On the basis of our earlier work, we believe income is related to education. But we know that education is not the only factor influencing income. Another factor is undoubtedly seniority. In most occupations, the longer one is on the job, the more money one makes. This seems likely to be so in Riverside city government. Therefore, our explanation for income differences should be improved if we revise our bivariate regression model to this multiple regression model:

$$Y = a_0 + b_1X_1 + b_2X_2 + e.$$

where Y = income (in dollars), X_1 = education (in years), X_2 = seniority (in years), e = error. The least squares estimates for the parameters are as follows:

$$\hat{Y} = 5666 + 432X_1 + 281X_2.$$

Interpreting the Parameter Estimates

The interpretation of the intercept, which merely extends the bivariate case, need not detain us: a_0 = the average value of Y when each independent variable equals zero. The interpretation of the slope, however, requires more attention: b_k = the average change in Y associated with a unit change in X_k, *when the other independent variables are held constant.* By this means of control, we are able to separate out the effect of X_k itself, free of any distorting influences from the other independent variables. Such a slope is called a *partial slope*, or *partial regression coefficient.* In the above Riverside example, partial slope b_2 estimates that a one-year increase in seniority is associated with an average income rise of $281, even assuming the employee's amount of education remains constant. In other words, a city worker can expect this annual salary increment, independent of any personal effort at educational improvement. Nevertheless, according to b_1, acquiring an additional year of schooling would add to an employee's income, regardless of the years of seniority accumulated. That is, an extra year of education will augment income an average of $432, beyond the benefits that come from seniority.

To appreciate fully the interpretation of the partial slope, one must grasp how multiple regression "holds constant" the other independent variables. First, it involves *statistical control*, rather than *experimental control.* For instance, in our Riverside study, if we were able to exercise experimental control, we might hold everyone's education at a constant value, say 10 years, and then record the effect on income of assigning respondents different amounts of seniority. To assess the effect of education on income, a similar experiment could be carried out. If such manipulation were possible, we could analyze the effects of seniority and education, respectively, by running two separate bivariate regressions, one on each experiment. However, since such experimental control is out of the question, we have to rely on the statistical control multiple regression provides. We can show how this statistical control operates to separate the effect of one independent variable from the others by examining the formula for a partial slope.

50

We confine ourselves to the following three-variable model, the results of which are generalizable:

$$Y = a_0 + b_1X_1 + b_2X_2 + e.$$

Let us explicate the b_1 estimation. Assuming $r_{12} \neq 0$, each independent variable can be accounted for, at least in part, by the other independent variables. That is, for example, X_1 can be written as a linear function of X_2,

$$X_1 = c_1 + c_2X_2 + u.$$

Supposing X_1 is not perfectly predicted by X_2, there is error, u. Hence, the observed X_1 can be expressed as the predicted X_1, plus error:

$$X_1 = \hat{X}_1 + u,$$

where $\hat{X}_1 = c_1 + c_2X_2$. The error, u, is the portion of X_1 which the other independent variable, X_2, cannot explain,

$$u = X_1 - \hat{X}_1.$$

This component, u, thus represents a part of X_1 which is completely separate from X_2.

By the same steps, we can also isolate the portion of Y which is linearly independent of X_2:

$$Y = d_1 + d_2X_2 + v$$
$$= (d_1 + d_2X_2) + v$$
$$Y = \hat{Y} + v.$$

The error, v, is that portion of Y which cannot be accounted for by X_2,

$$v = Y - \hat{Y}.$$

This component, v, then, is that part of Y which is unrelated to X_2.

These two error components, u and v, are joined in the following formula for b_1:

$$b_1 = \frac{\Sigma(u)(v)}{\Sigma u^2} = \frac{\Sigma(X_1 - \hat{X}_1)(Y - \hat{Y})}{\Sigma(X_1 - \hat{X}_1)^2}.$$

In words, b_1 *is determined by* X_1 *and* Y *values that have been freed of any linear influence from* X_2. In this way, the effect of X_1 is separated from the effect of X_2. The formula, generally applicable for any partial slope, should be familiar, for we saw a special version of it in the bivariate case, where

$$b = \frac{\Sigma(X - \bar{X})(Y - \bar{Y})}{\Sigma(X - \bar{X})^2}.$$

While the statistical control of multiple regression is weaker than experimental control, it still has great value. The careful introduction of additional variables into an equation permits greater confidence in our findings. For instance, the bivariate regression model of the Riverside study suggested that education is a determinant of income. However, this conclusion is open to challenge. That apparent bivariate relationship could be spurious, a product of the common influence of another variable on education and income. For example, an antagonist might argue that the observed bivariate relationship is actually caused by senority, for those with more years on the job are those with more education, as well as higher pay. An implication is that if seniority were "held constant," education would be exposed as having no effect on income. Multiple regression permits us to test this hypothesis of spuriousness. From the above least squares estimates, we discovered that education still has an apparent effect, even after taking the influence of seniority into account. Hence, through actually bringing this third variable into the equation, we are able to rule out an hypothesis of spuriousness, and thereby strengthen our belief that education affects income.

Confidence Intervals and Significance Tests

The procedure for confidence intervals and significance tests carries over from the bivariate case. Suppose we wish to know whether the partial slope estimate, b_1, from our three-variable equation for the Riverside study, is significantly different from zero. Again, we confront the null hypothesis, which says there is no relationship in the population, and the alternative hypothesis, which says there is a relationship in the population. Let us construct a two-tailed, 95% confidence interval around this partial slope estimate, in order to test these hypotheses:

$$(b_1 \pm t_{n-3;.975} s_b).$$

Note that the only difference between this formula and the bivariate formula is the number of degrees of freedom. Here, we have one less degree of freedom, $(n-3)$ instead of $(n-2)$, because we have one more independent variable. In general, the degrees of freedom of the t variable equal $(n-k-1)$, where n = sample size and k = number of independent variables. Applying the formula,

$$(432 \pm t_{29;.975} s_b) = 432 \pm 2.045\ (144) = (432 \pm 294).$$

The probability is .95 that the value of the partial slope in the population is between \$138 and \$726. Because the value of zero is not captured within this band, we reject the null hypothesis. We state that the partial slope estimate, b_1, is significantly different from zero, at the .05 level.

A second approach to the significance testing of b_1 would be examination of the t ratio,

$$b_1/s_{b_1} = 432/144 = 3.01.$$

We observe that the value of this t ratio exceeds the t distribution value, $t_{n-3;\ 975}$. That is,

$$3.01 > 2.045.$$

Therefore, we conclude that b_1 is statistically significant at the .05 level.

The most efficient means of significance testing is to use the rule of thumb, which claims statistical significance at the .05 level, two-tailed, for any coefficient whose t ratio exceeds 2 in absolute value. Below is the three-variable Riverside equation, with the t ratios in parentheses:

$$\hat{Y} = 5666 + 432X_1 + 281X_2.$$
$$(4.22)\quad (3.01)\qquad (3.04)$$

An examination of these t ratios, with this rule of thumb in mind, instantly reveals that all the parameter estimates of the model (a_0, b_1, b_2) are significant at the .05 level.

The R^2

To assess the goodness of fit of a multiple regression equation, we employ the R^2, now referred to as the *coefficient of multiple determination*. Once again,

$$R^2 = \frac{\Sigma(\hat{Y} - \bar{Y})^2}{\Sigma(Y - \bar{Y})^2} = \frac{\text{regression (explained) sum of squares}}{\text{total sum of squares}}.$$

The R^2 for a multiple regression equation indicates the proportion of variation in Y "explained" by all the independent variables. In the above three-variable Riverside model, $R^2 = .67$, indicating that education and seniority together account for 67% of the variance in income. This multiple regression model clearly provides a more powerful explanation for income differences than the bivariate regression model, where $R^2 = .56$.

Obviously, it is desirable to have a high R^2, for it implies a more complete explanation of the phenomenon under study. Nevertheless, if a higher R^2 were the only goal, then one could simply add independent variables to the equation. That is, an additional independent variable cannot lower the R^2, and is virtually certain to increase it at least somewhat. In fact, if independent variables are added until their number equals n−1, then $R^2 = 1.0$. This "perfect" explanation is of course nonsense, and amounts to no more than a mathematical necessity, which occurs because the degrees of freedom have been exhausted. In sum, rather than entering variables primarily to enhance R^2, the analyst must be guided by theoretical considerations in deciding which variables to include.

Predicting Y

A multiple regression equation is used for prediction as well as explanation. Let us predict the income of a Riverside city employee who has 10 years of education and has been on the job 5 years:

$$\hat{Y} = 5666 + 432X_1 + 281X_2$$
$$= 5666 + 432(10) + 281(5)$$
$$= 5666 + 4320 + 1405$$
$$\hat{Y} = 11,391.$$

In order to get some notion of the accuracy of this prediction, we can construct a confidence interval around it, utilizing the standard error of estimate of Y, s_e:

$$(\hat{Y} \pm 2s_e) = \hat{Y} \pm 2(2529) = 11,391 \pm 5058.$$

This confidence interval indicates there is a 95% chance that a municipal employee with 10 years of education and 5 years of seniority will earn between $6333 and $16449. While this prediction is more accurate than that generated by the bivariate regression equation, it is still far from precise.

The model is even less useful for forecasting beyond its range of experience. Certainly, we could plug in any values for X_1 and X_2 and produce a prediction for Y. However, the worth of the forecast diminishes as these X_1 and X_2 values depart from the actual range of variable values in the data. For instance, it would be risky to predict the income of a city worker with two years of education and 35 years of seniority, for no one in the data-set registered such extreme scores. Possibly, at such extreme values, the linearity of the relationships would no longer exist. Then, any prediction based on our linear model would be quite wide of the mark.

The Possibility of Interaction Effects

Thus far, we have assumed that effects are *additive*. That is, Y is determined, in part, by X_1 *plus* X_2, not X_1 *times* X_2. This additivity assumption dominates applied regression analysis and is frequently justified. However, it is not a necessary assumption. Let us explore an example.

We have mentioned the variable of sex of respondent as a candidate for inclusion in the Riverside income equation. The question is, should the sex variable enter additively or as an interaction. It might be argued that sex is involved interactively with education. In general, an *interaction effect* exists when the impact of one independent variable depends on the value of another independent variable. Specifically, perhaps the effect of education is dependent on the sex of the employee, with education yielding a greater financial return for men.

Formally, this particular interaction model is as follows (we ignore the seniority variable for the moment):

$$\hat{Y} = a_0 + b_1 X_1 + b_2 (X_1 X_2) + e,$$

where Y = income (in dollars); X_1 = education (in years); X_2 = sex of respondent (0 = female, 1 = male); $X_1 X_2$ = an interaction variable created by multiplying X_1 times X_2. The least squares estimates for this model are,

$$\hat{Y} = 5837 + 556 X_1 + 202 (X_1 X_2) \qquad R^2 = .65,$$
$$\quad (4.20) \quad (4.44) \quad (2.70)$$

where the figures in parentheses are t ratios. These results indicate that education, while increasing the income of both sexes, provides a greater income increase for men. This becomes clearer when we separate out the prediction equations for men and women.

Prediction equation for women:

$$\hat{Y} = a_0 + b_1 X_1 + b_2 X_1(0)$$
$$= a_0 + b_1 X_1$$
$$\hat{Y} = 5837 + 556 X_1.$$

Prediction equation for men:

$$\hat{Y} = a_0 + b_1 X_1 + b_2 X_1(1)$$
$$= a_0 + (b_1 + b_2) X_1$$
$$\hat{Y} = 5837 + 758 X_1.$$

We observe that, for men, the slope for the education variable is greater. Further, this slope difference is statistically significant (see the t ratio for b_2).

The rival, strictly additive, model, is,

$$Y = a_0 + b_1 X_1 + b_2 X_2 + e,$$

where the variables are defined as before. Estimating this model yields,

$$\hat{Y} = 4995 + 633 X_1 + 2555 X_2 \qquad R^2 = .65,$$
$$(3.64) \quad (5.54) \quad (2.60)$$

where the values in parentheses are t ratios. These estimates suggest that education and sex have significant, independent effects on income.

The data are congruent with both the interaction model and the additive model. The coefficients are all statistically significant, and the R^2 is the same in both. Which model is correct? The answer must base itself on theoretical considerations and prior research, since the empirical evidence does not permit us to decide between them. The additive model seems more in keeping with a "discrimination" theory of income determination; that is, other things being equal, society pays women less solely because they are women. The interaction model appears to square better with an "individual failure" theory of income determination; that is, women are paid less because they are less able to make education work to their advantage. On the basis of prior theorizing and research, I favor the "discrimination" interpretation and therefore choose to allow the sex variable to enter the larger income equation additively. (A resolution of the two models might come from estimation of an equation which allows sex to have additive and interactive effects:

$$Y = a_0 + b_1X_1 + b_2X_2 + b_3(X_1X_2) + e.$$

Unfortunately, the estimates from this model are made unreliable by severe multicollinearity, a problem not uncommon with interaction models. We consider multicollinearity at length below.)

A Four-Variable Model: Overcoming Specification Error

Incorporating the sex variable additively into our model for income differences in Riverside leads to the following equation:

$$Y = a_0 + b_1X_1 + b_2X_2 + b_3X_3 + e,$$

where Y = income (in dollars), X_1 = education (in years), X_2 = seniority (in years), X_3 = sex (0 = female, 1 = male), e = error. Theoretically, this four-variable model is much more complete than the earlier two-variable model. It asserts that income is a linear additive function of three factors: education, seniority, and sex.

Estimating this multiple regression model with least squares yields,

$$\hat{Y} = 5526 + 385X_1 + 247X_2 + 2140X_3$$
$$(4.44) \quad (2.86) \quad (2.84) \quad (2.40)$$

$$R^2 = .73 \quad n = 32 \quad s_e = 2344,$$

where the values in parentheses are the t ratios, R^2 = coefficient of multiple determination, n = sample size, s_e = standard error of estimate of Y.

These estimates tell us a good deal about what affects income in Riverside city government. The pay of a municipal employee is significantly influenced by years of education, amount of seniority, and sex. (Each t ratio exceeds 2, indicating statistical significance at the .05 level.) These three factors largely determine income differences within this population. In fact, almost three-quarters of the variation in income is explained by these variables (R^2 = .73). The differences caused are not inconsequential. For each year of education, $385 is added to income, on the average. An extra year of seniority contributes another $247. Male workers can expect $2140 more than females workers, even if the women have the same education and seniority. The cumulative impact of these variables can create sizable income disparities. For example, a male with a college education and 10 years seniority would expect to make $16,296; in contrast, a female with a high school degree and just starting work could only expect to earn $10,146.

Inclusion of relevant variables, that is, seniority and sex, beyond the education variable, has markedly diminished specification error, helping ensure that our estimates are BLU. (To refresh yourself on the meaning of specification error, review the discussion of assumptions in Chapter 2.) In particular, the estimate of the education coefficient, which equaled 732 in the bivariate model, has been sharply reduced. The comparable estimate in this four-variable model, $b_1 = 385$, indicates that the true impact of education is something like one-half that estimated in the original bivariate equation.

For certain models, it is fairly easy to detect the direction of bias resulting from the exclusion of a relevant variable. Suppose the real world is congruent with this model;

$$Y = a_0 + b_1 X_1 + b_2 X_2 + e \quad \text{(correct model)},$$

but we mistakenly estimate,

$$Y = a_0 + b_1 X_1 + e^* \quad \text{(incorrect model)},$$

where $e^* = (b_2 X_2 + e)$. By excluding X_2 from our estimation, we have committed specification error. Assuming that X_1 and X_2 are correlated, as they almost always are, the slope estimate, b_1, will be biased. This bias is inevitable, for the independent variable, X_1, and the error term, e^*, are correlated, thus violating an assumption necessary for regression to yield desirable estimators. (We see that $r_{x_1 e^*} \neq 0$, because $r_{x_1 x_2} \neq 0$, and X_2 is a component of e^*.) The direction of the bias of b_1 in the estimated model is determined by: (1) the sign of b_2 and (2) the sign of the correlation, r_{12}. If b_2 and r_{12} have the same sign, then the bias of b_1 is positive; if not, then the bias is negative.

It happens that the direction of bias in the somewhat more complicated Riverside case accords with these rules. As noted, the bias of b_1 in the bivariate equation of the Riverside study is positive, accepting the specification and estimation of the four-variable model. The presence of this positive bias follows the above guidelines: (1) the sign of b_2 (and b_3) is positive and (2) the sign of r_{12} (and r_{13}) is positive; therefore, the bivariate estimate of b_1 must be biased upward. Part of the variance in Y that X_1 is accounting for should be explained by X_2 and X_3, but these variables are not in the equation. Thus, some of the impact of X_2 and X_3 on Y is erroneously assigned to X_1.

The formulation of rules for the detection of bias implies that it is possible to predict the consequences of a given specification error. For instance, the analyst is able to foresee the direction of bias coming from the exclusion of a certain variable. With simpler models, such as those treated here, such insight might be attainable. However, for models which include several variables, and face several candidates for inclusion, the direction of bias is not readily foreseeable. In this more complex situation, the analyst is better served by immediate attention to proper specification of the model.

The Multicollinearity Problem

For multiple regression to produce the "best linear unbiased estimates," it must meet the bivariate regression assumptions, plus one additional assumption: the absence of *perfect multicollinearity*. That is, none of the independent variables is perfectly correlated with another independent variable or linear combination of other independent variables. For example, with the following multiple regression model,

$$Y = a_0 + b_1X_1 + b_2X_2 + e,$$

perfect multicollinearity would exist if,

$$X_2 = c_0 + c_1X_1,$$

for X_2 is a perfect linear function of X_1 (that is, $R^2 = 1.0$). When perfect multicollinearity exists, it is impossible to arrive at a unique solution for the least squares parameter estimates. Any effort to calculate the partial regression coefficients, by computer or by hand, will fail. Thus, the presence of perfect multicollinearity is immediately detectable. Further, in practice, it is obviously quite unlikely to occur. However, *high multicollinearity* commonly perplexes the users of multiple regression.

With nonexperimental social science data, the independent variables are virtually always intercorrelated, that is, multicollinear. When this condition becomes extreme, serious estimation problems often arise. The general difficulty is that parameter estimates become unreliable. The magnitude of the partial slope estimate in the present sample may differ considerably from its magnitude in the next sample. Hence, we have little confidence that a particular slope estimate accurately reflects the impact of X on Y in the population. Obviously, because of such imprecision, this partial slope estimate cannot be usefully compared to other partial slope estimates in the equation, in order to arrive at a judgment of the relative effects of the independent variables. Finally, an estimated regression coefficient may be so unstable that it fails to achieve statistical significance, even though X is actually associated with Y in the population.

High multicollinearity creates these estimation problems because it produces large variances for the slope estimates and, consequently, large standard errors. Recalling the formula for a confidence interval (95%, two-tailed),

$$(b \pm t_{n-k-1;.975} s_b),$$

we recognize that a larger standard error, s_b, will widen the range of values that b might take on. Reviewing the formula for the t ratio,

$$b / s_b,$$

we observe that a larger s_b makes it more difficult to achieve statistical significance (e.g., more difficult to exceed the value of 2, which indicates statistical significance at the .05 level, two-tailed).

We can see how large variances occur with high multicollinearity by examining this variance formula,

$$\text{variance } b_i = s_{b_i}^2 = s_u^2 / v_i^2$$

where s_u^2 is the variance of the error term in the multiple regression model, and v_i^2 is the squared residual from the regression of the ith independent variable, X_i, on the rest of the independent variables in the model. Hence,

$$v_i = X_i - \hat{X}_i.$$

If these other independent variables are highly predictive of X_i, then X_i and $\hat{X}_i$ will be very close in value, and so v_i will be small. Therefore, the denominator in the above variance formula will be small, yielding a large variance estimate for b_i.

Of course, when analysts find a partial regression coefficient is statistically insignificant, they cannot simply dismiss the result on grounds of high multicollinearity. Before such a claim can be made, high multicollinearity must be demonstrated. Let us first look at common symptoms of high multicollinearity, which may alert the researcher to the problem. Then, we will proceed to a technique for diagnosis. One rather sure symptom of high multicollinearity is a substantial R^2 for the equation, but statistically insignificant coefficients. A second, weaker, signal is regression coefficients which change greatly in value when independent variables are dropped or added to the equation. A third, still less sure, set of symptoms involves suspicion about the magnitudes of the coefficients. A coefficient may be regarded as unexpectedly large (small), either in itself, or relative to another coefficient in the equation. It may even be so large (or small) as to be rejected as nonsensical. A fourth alert is a coefficient with the "wrong" sign. Obviously, this last symptom is feeble, for knowledge of the "right" sign is often lacking.

The above symptoms might provide the watchful analyst hints of a multicollinearity problem. However, by themselves, they cannot establish that the problem exists. For diagnosis, we must look directly at the inter-correlation of the independent variables. A frequent practice is to examine the bivariate correlations among the independent variables, looking for coefficients of about .8, or larger. Then, if none is found, one goes on to conclude that multicollinearity is not a problem. While suggestive, this approach is unsatisfactory, for it fails to take into account the relationship of an independent variable with *all* the other independent variables. It is possible, for instance, to find no large bivariate correlations, although one of the independent variables is a nearly perfect linear combination of the remaining independent variables. This possibility points to the preferred method of assessing multicollinearity: *Regress each independent variable on all the other independent variables.* When any of the R^2 from these equations is near 1.0, there is high multicollinearity. In fact, the largest of these R^2 serves as an indicator of the amount of multicollinearity which exists.

Let us apply what we have learned about multicollinearity to the four-variable Riverside model,

$$Y = a_0 + b_1X_1 + b_2X_2 + b_3X_3 + e,$$

where Y = income, X_1 = education, X_2 = seniority, X_3 = sex, e = error. The estimates for this model, which we have already examined, reveal no symptoms of a multicollinearity problem. That is, the coefficients are all significant, and their signs and magnitudes are reasonable. Therefore, we would anticipate that the above multicollinearity test would produce $R^2_{x_i}$ far from unity. Regressing each independent variable on all the others yields,

$$\hat{X}_1 = 7.02 + .42X_2 + .96X_3 \qquad R^2 = .49$$
$$\hat{X}_2 = -2.15 + 1.00X_1 + 1.68X_3 \qquad R^2 = .49$$
$$\hat{X}_3 = .066 + .022X_1 + .016X_2 \qquad R^2 = .14.$$

These $R^2_{x_i}$ show that these independent variables are intercorrelated in the Riverside sample, as we would expect with data of this type. But, we observe that the largest coefficient of multiple determination, $R^2 = .49$, lies a good distance from 1.0. Our conclusion is that multicollinearity is not a problem for the partial slope estimates in the Riverside multiple regression model.

The results do not always turn out so well. What can we do if high multi-collinearity is detected? Unfortunately, none of the possible solutions is wholly satisfactory. In general, we must make the best of a bad situation. The standard prescription is to increase our information by enlarging the sample. As noted in an earlier chapter, the bigger the sample size, the greater the chance of finding statistical significance, other things being equal. Realistically, however, the researcher is usually unable to increase the sample. Also, multicollinearity may be severe enough that even a large n will not provide much relief.

Assuming the sample size is fixed, other strategies have to be imple-mented. One is to combine those independent variables that are highly intercorrelated into a single indicator. If this approach makes conceptual sense, then it can work well. Suppose, for example, a model which explains political participation (Y) as a function of income (X_1), race (X_2), radio listening (X_3), television watching (X_4), and newspaper reading (X_5). On the one hand, it seems sensible to combine the highly intercorrelated variables (X_3, X_4, X_5) into an index of media involvement. On the other hand, it is not sensible to combine the income and race variables, even if they are highly related.

Suppose our variables are "apples and oranges," making it impractical to combine them. In the face of high multicollinearity, we cannot reliably separate the effects of the involved variables. Still, the equation may have value if its use is restricted to prediction. That is, it might be employed to predict Y for a given set of values on *all* the X's (e.g., when $X_1 = 2$, $X_2 = 4$, . . . $X_k = 3$), but not to interpret the independent effect on Y of a change in the value of a *single* X. Usually, this prediction strategy is uninteresting, for the goal is generally explanation, in which we talk about the impact of a particular X on Y.

A last technique for combatting multicollinearity is to discard the offending variable(s). Let us explore an example. Suppose we specify the following elementary multiple regression model,

$$Y = a_0 + b_1X_1 + b_2X_2 + e \qquad \text{Model I.}$$

Lamentably, however, we find that X_1 and X_2 are so highly related ($r_{12} = .9$), that the least squares estimates are unable reliably to assess the effect of either. An alternative is to drop one of the variables, say X_2, from the equation, and simply estimate this model:

$$Y = a_0 + b_1 X_1 + e^* \qquad \text{Model II.}$$

A major problem with this procedure, of course, is its willful commission of specification error. Assuming Model I is the correct explanatory model, we know the estimate for b_1 in Model II will be biased. A revision which makes this technique somewhat more acceptable is to estimate yet another equation, now discarding the other offending variable (X_1),

$$Y = a_0 + b_2 X_2 + e^{**} \qquad \text{Model III.}$$

If the Model II and Model III estimates are evaluated, along with those of Model I, then the damage done by the specification error can be more fully assessed.

High Multicollinearity: An Example

In order to grasp more completely the influences of high multicollinearity, it is helpful to explore a real data example. First, we present research findings reported by sociologist Gino Germani (1973). Then, we examine these findings with an eye to the multicollinearity issue.[4] Germani wishes to explain the vote support Juan Peron garnered in the 1946 presidential election in Argentina. His special interest is in assessing the backing Peron received from workers and internal migrants. To do so, he formulates a multiple regression model, arriving at the following estimates,

$$\hat{Y} = .52 + .18X_1 - .10X_2 - .57X_3 - 3.57X_4 + .29^*X_5$$
$$\quad\;\; (.43) \quad (.41) \quad (.43) \quad (2.54) \quad (.07)$$

$$R^2 = .24 \qquad n = 181 \qquad s_e = .11,$$

where Y = the percentage of the county's 1946 presidential vote going to Peron; X_1 = urban blue-collar workers (as a percentage of the economically active population in the county); X_2 = rural blue-collar workers (as a percentage of the economically active population in the county); X_3 = urban white-collar workers (as a percentage of the economically active population in the county); X_4 = rural white-collar workers (as a percentage of the economically active population in the county); X_5 = internal migrants (as a percentage of Argentinian-born males); the figures in parentheses are the standard errors of the slope estimates; the asterisk, *, indicates a coefficient statistically significant at the .05 level, two-tailed;

R^2 = coefficient of multiple determination; n = 181 counties that contained a city of at least 5000 people; s_e = the standard error of estimate of Y.

These results suggest that only the presence of internal migrants significantly affected Peron support. We are pushed to the conclusion that the workers were not an influential factor in the election of Juan Peron. Such a conclusion becomes much less certain when we inspect the multicollinearity in the data. Let us diagnose the level of multicollinearity by regressing each independent variable on the remaining independent variables. This yields the following $R^2_{x_i}$, in order of magnitude: $R^2_{x_2}$ = .99, $R^2_{x_3}$ = .98, $R^2_{x_1}$ = .98, $R^2_{x_4}$ = .75, $R^2_{x_5}$ = .32.

Obviously, extreme multicollinearity is present. How might it be corrected? Further observations cannot be gathered. It is not sensible to combine any of the variables into an index. The purpose of the equation is not prediction. (If it were, the low R^2_y would inhibit it.) We are left with the strategy of discarding offending variables. An examination of the $R^2_{x_i}$ shows that the largest is $R^2_{x_2}$. The variable, X_2, is an almost perfect linear function of all the other independent variables (X_1, X_3, X_4, X_5). Suppose we remove X_2 from the equation and reestimate:

$$\hat{Y} = .42 + .28^*X_1 - .47^*X_3 - 3.07^*X_4 + .30^*X_5$$
$$\phantom{\hat{Y} = .42 + } (.07) \quad\quad (.10) \quad\quad (1.41) \quad\quad (.07)$$

$$R^2 = .24 \quad\quad n = 181 \quad\quad s_e = .11,$$

where definitions are the same as above.

According to these new estimates, *all* the variables have a statistically significant impact. Contrary to the earlier conclusion, workers do appear to have contributed to the election of Peron. How reliable are these new estimates? One check is to recalculate the level of multicollinearity. Regressing each independent variable on the remaining variables in the revised equation yields, $R^2_{x_3}$ = .38, $R^2_{x_5}$ = .30, $R^2_{x_1}$ = .29, $R^2_{x_4}$ = .20. We observe that all of these $R^2_{x_i}$ are quite far from unity, indicating that multicollinearity has ceased to be problematic. The revised parameter estimates would appear much more reliable than the contrary ones generated with the offending X_2 in the equation. Hopefully, this rather dramatic example brings home the perils of high multicollinearity.

The Relative Importance of the Independent Variables

We sometimes want to evaluate the relative importance of the independent variables in determining Y. An obvious procedure is to compare the magnitudes of the partial slopes. However, this effort is often thwarted by the different measurement units and variances of the variables. Suppose, for example, the following multiple regression equation predicting dollars contributed to political campaigns as a function of an individual's age and income,

$$\hat{Y} = 8 + 2X_1 + .010X_2,$$

where Y = campaign contributions (in dollars), X_1 = age (in years), X_2 = income (in dollars).

The relative influence of income and age on campaign contributions is difficult to assess, for the measurement units are not comparable, that is, dollars versus years. One solution is to *standardize* the variables, re-estimate, and evaluate the new coefficients. (Some computing routines for regression, such as that of SPSS, automatically provide the standardized coefficients along with the unstandardized coefficients.) Any variable is standardized by converting its scores into standard deviation units from the mean. For the above variables, then,

$$Y^* = \frac{Y - \bar{Y}}{s_y}, \quad X_1^* = \frac{X_1 - \bar{X}_1}{s_{x_1}}, \quad X_2^* = \frac{X_2 - \bar{X}_2}{s_{x_2}},$$

where the asterisk, *, indicates the variable is standardized.

Reformulating the model with these variables yields,

$$\hat{Y}^* = \beta_1 X_1^* + \beta_2 X_2^*.$$

(Note that standardization forces the intercept to zero.) The standardized partial slope is often designated with "β," and referred to as a *beta weight*, or *beta coefficient*. (Do not confuse this β with the symbol for the population slope.)

The beta weight corrects the unstandardized partial slope by the ratio of the standard deviation of the independent variable to the standard deviation of the dependent variable:

$$\beta_i = b_i \frac{s_{x_i}}{s_y}.$$

In the special case of the bivariate regression model, the beta weight equals the simple correlation between the two variables. That is, assuming the model,

$$Y = a + bX + e,$$

then,

$$\beta = b \ \frac{s_x}{s_y} = r.$$

However, this equality does not hold for a multiple regression model. (Only in the unique circumstance of *no* multicollinearity would $\beta = r$ with a multiple regression model.)

The standardized partial slope estimate, or beta weight, indicates *the average standard deviation change in Y associated with a standard deviation change in X, when the other independent variables are held constant.* Suppose the beta weights for the above campaign contribution equation are as follows:

$$\hat{Y}^* = .15X_1^* + .45X_2^*.$$

For example, $\beta_2 = .45$ says that a one standard deviation change in income is associated with a .45 standard deviation change in campaign contributions, on the average, with age held constant. Let us consider the meaning of this interpretation more fully. Assuming X_2 is normally distributed, then a one standard deviation income rise for persons at, say, the mean income would move them into a high income bracket, above which only about 16% of the population resided. We see that this strong manipulation of X does not result in as strong a response in Y, for β_2 is far from unity. Still, campaign contributions do tend to climb by almost one-half of a standard deviation. In contrast, a considerable advance in age (a full one standard deviation increase) elicits a very modest increment in contributions (only .15 of a standard deviation). We conclude that the impact of income, as measured in standard deviation units, is greater than the impact of age, likewise measured. Indeed, it seems that the effect of income on campaign contributions is three times that of age ($.45/.15 = 3$).

The ability of standardization to assure the comparability of measurement units guarantees its appeal, when the analyst is interested in the relative effects of the independent variables. However, difficulties can arise if one wishes to make comparisons across samples. This is because, in estimating the same equation across samples, the value of the beta weight, unlike the value of the unstandardized slope, can change merely because the variance of X changes. In fact, the larger (smaller) the variance in X, the larger (smaller) the beta weight, other things being equal. (To understand this, consider again the beta weight formula,

$$\beta_i = b_i \ \frac{s_{x_i}}{s_y}.$$

We see that, as s_{x_1}, the numerator of the fraction, increases, the magnitude of β_i must necessarily increase.)

As an example, suppose that the above campaign contributions model was developed from a U.S. sample, and we wished to test it for another Western democracy, say Sweden. Our beta weights from this hypothetical sample of the Swedish electorate might be,

$$\hat{Y}^* = .18X_1^* + .22X_2^*,$$

where the variables are defined as above. Comparing β_2 (United States) = .45 to β_2 (Sweden) = .22, we are tempted to conclude that the effect of income in Sweden is about one-half its effect in the United States. However, this inference may well be wrong, given that the standard deviation of of income in the United States is greater than the standard deviation of income in Sweden. That is, the wider spread of incomes in the United States may be masking the more equal effect a unit income change actually has in both countries, that is, b_2 (United States) $\cong$ b_2 (Sweden). To test for this possibility, we must of course examine the unstandardized partial slopes, which we suppose to be the following:

$$\hat{Y} = 9 + 1.7X_1 + .012X_2.$$

When these unstandardized Swedish results are compared to the unstandardized United States results, they suggest that, in reality, the effect of income on campaign contributions is essentially the same in both countries (.010 $\cong$ .012). In general, when the variance in X diverges from one sample to the next, it is preferable to base any cross-sample comparisons of effect on the unstandardized partial slopes.

Extending the Regression Model: Dummy Variables

Regression analysis encourages the use of variables whose amounts can be measured with numeric precision, that is, *interval variables*. A classic example of such a variable is income. Individuals can be ordered numerically according to their quantity of income, from the lowest to the highest. Thus, we can say that John's income of $12,000 is larger than Bill's income of $6,000; in fact, it is exactly twice as large. Of course, not all variables are measured at a level which allows such precise comparison. Nevertheless, these noninterval variables are candidates for incorporation into a regression framework, through the employment of *dummy variables*.

Many noninterval variables can be considered *dichotomies*, e.g., sex (male, female), race (Black, White), marital status (single, married). Dichotomous independent variables do not cause the regression estimates to lose any of their desirable properties. Because they have two categories, they manage to "trick" least squares, entering the equation as an interval variable with just two values. It is useful to examine how such "dummy" variables work. Suppose we argue that a person's income is predicted by race in this bivariate regression,

$$\hat{Y} = a + bX,$$

where Y = income, X = race (0 = Black, 1 = White). If X = 0, then

$$\hat{Y} = a,$$

the prediction of the mean income for Blacks. If X = 1, then

$$\hat{Y} = a + b,$$

the prediction of the mean income for Whites. Therefore, the slope estimate, b, indicates the difference between the mean incomes of Blacks and Whites. As always, the t ratio of b measures its statistical significance. We have already observed such a dummy variable in action, in the four-variable Riverside equation, which included sex as an independent variable (0 = female, 1 = male). There, the partial regression coefficient, b_3, reports the difference in average income between men and women, after the influences of education and seniority have been accounted for. As noted, this difference is statistically and substantively signficant.

Obviously, not all noninterval variables are dichotomous. Noninterval variables with multiple categories are of two basic types: *ordinal* and *nominal*. With an ordinal variable, cases can be ordered in terms of amount, but not with numeric precision. Attitudinal variables are commonly of this kind. For example, in a survey of the electorate, respondents may be asked to evaluate their political interest, ranking themselves as "not interested," "somewhat interested," or "very interested." We can say that Respondent A, who chooses "very interested," is more interested in politics than Respondent B, who selects "not interested," but we cannot say numerically how much more. Ordinal variables, then, only admit of a ranking from "less to more." The categories of a nominal variable, in contrast, cannot be so ordered. The variable of religious affiliation is a good example. The categories of Protestant, Catholic, or Jew represent personal attributes which yield no meaningful ranking.

Noninterval variables with multiple categories, whether ordinal or nominal, can be incorporated into the multiple regression model through the dummy variable technique. Let us explore an example. Suppose the dollars an individual contributes to a political campaign are a function of the above-mentioned ordinal variable, political interest. Then, a correct model would be

$$Y = a_0 + b_1X_1 + b_2X_2 + e,$$

where Y = campaign contributions (in dollars); X_1 = a dummy variable, scored 1 if "somewhat interested," 0 if otherwise; X_2 = a dummy variable, scored 1 if "very interested," 0 if otherwise; e = error.

Observe that there are only *two* dummy variables to represent the trichotomous variable of political interest. If there were three dummy variables, then the parameters could not be uniquely estimated. That is, a third dummy, X_3 (scored 1 if "not interested," 0 if otherwise), would be an exact linear function of the others, X_1 and X_2. (Consider that when the score of any respondent on X_1 and X_2 is known, it would always be possible to predict his or her X_3 score. For example, if a respondent has values of 0 on X_1 and 0 on X_2, then he or she is necessarily "not interested" in politics, and would score 1 on X_3.) This describes a situation of perfect multicollinearity, in which estimation cannot proceed. To avoid such a trap, which is easy to fall into, we memorize this rule: *When a noninterval variable has G categories, use G − 1 dummy variables to represent it.*

A question now arises as to how to estimate the campaign contributions of this excluded group, those who responded "not interested." Their average campaign contribution is estimated by the intercept of the equation. That is, for someone who is "not interested," the prediction equation reduces to,

$$\hat{Y} = a_0 + b_1X_1 + b_2X_2$$
$$= a_0 + b_1(0) + b_2(0)$$
$$\hat{Y} = a_0.$$

Thus, the intercept estimates the average campaign contribution of someone who is "not interested" in politics.

This estimated contribution, a_0, for the "not interested" category serves as a base for comparing the effects of the other categories of political interest. The prediction equation for someone in the category, "somewhat interested," reduces to

$$\hat{Y} = a_0 + b_1X_1 + b_2X_2$$
$$= a_0 + b_1(1) + b_2(0)$$
$$\hat{Y} = a_0 + b_1.$$

Hence, the partial slope estimate, b_1, indicates the difference in mean campaign contributions between those "somewhat interested" and those "not interested," that is, $(a_0 + b_1) - a_0 = b_1$.

For the last category, "very interested," the prediction equation reduces to

$$\hat{Y} = a_0 + b_1 X_1 + b_2 X_2$$
$$= a_0 + b_1(0) + b_2(1)$$
$$\hat{Y} = a_0 + b_2.$$

Thus, the partial slope estimate, b_2, points out the difference in average campaign contributions between the "very interested" and the "not interested." Given the hypothesis that heightened political interest increases campaign contributions, we would expect that $b_2 > b_1$.

A data example will increase our appreciation of the utility of dummy variables. Suppose, with the Riverside study, it occurs to us that the income received from working for city government might be determined in part by the employee's political party affiliation (Democrat, Republican, or independent). In that case, the proper specification of the model becomes,

$$Y = a_0 + b_1 X_1 + b_2 X_2 + b_3 X_3 + b_4 X_4 + b_5 X_5 + e,$$

where Y = income; X_1 = education; X_2 = seniority; X_3 = sex; X_4 = a dummy variable scored 1 if independent, 0 otherwise; X_5 = a dummy variable scored 1 if Republican, 0 otherwise; e = error.

The variable, political party, has three categories. Thus, applying the $G - 1$ rule, we had to formulate $3 - 1 = 2$ dummy variables. We chose to construct one for independents (X_4) and one for Republicans (X_5), which left Democrats as the base category. The selection of a base category is entirely up to the analyst. Here, we selected Democrats as the standard for comparison because we guessed they would have the lowest income, with independents and Republicans having successively higher incomes.

Least squares yields the following parameter estimates,

$$\hat{Y} = 5496 + 382X_1 + 250X_2 + 2134X_3 - 572X_4 + 386X_5$$
$$\phantom{\hat{Y} = 5496 + } (3.90) \quad (2.74) \quad (2.78) \quad (2.33) \quad (-.48) \quad (.41)$$

$$R^2 = .73 \qquad n = 32 \qquad s_e = 2403,$$

where the variables are defined as above, the values in parentheses are t ratios, the R^2 = the coefficient of multiple determination, n = sample size, s_e = standard error of estimate for Y.

First, we note that the estimates from our prior specification remain virtually unchanged. Further, from the t ratio, we see that the average income of independents is not significantly different (.05 level) from the average income of Democrats, once the effects of education, seniority, and sex are controlled. (Put another way, b_4 does not add significantly to the intercept, a_0.) Likewise, the average income of Republicans is found not to differ significantly from that of the Democrats. We must conclude that, contrary to our expectation, political party affiliation does not influence the income of Riverside municipal employees. Our original four-variable model remains the preferred specification.

Through use of the dummy variable technique, the inclusion into our multiple regression equation of the noninterval variable, political party, poses no problem. Some researchers would argue that this variable could be inserted into our regression equation directly, bypassing the dummy variable route. The argument is that an ordinal variable is a candidate for regression, even though the distances between the categories are not exactly equal. This is a controversial point of view. In brief, the advocates' primary defense is that, in practice, the conclusions are usually equivalent to those generated by more correct techniques (i.e., the application of dummy variable regression or ordinal-level statistics). A secondary argument is that multiple regression analysis is so powerful, compared to ordinal-level techniques, that the risk of error is acceptable. We cannot resolve this debate here. However, we can provide a practical test by incorporating political party into the Riverside equation as an ordinal variable.

At first blush, political party affiliation may appear as strictly nominal. Nevertheless, political scientists commonly treat it as ordinal. We can say, for example, that an independent is "more Republican" than a Democrat, who is "least Republican" of all. Hence, we can order the categories in terms of their "distance" from Republicans. This order is indicated in the following numeric code, Democrat = 0, independent = 1, Republican = 2, which ranks the categories along this dimension of "Republicanism." This code provides each respondent a score on a political party variable, X_4, which we now enter into the Riverside equation. Least squares yields the following estimates,

$$\hat{Y} = 5314 + 392X_1 + 243X_2 + 2137X_3 + 186X_4$$
$$\quad\quad (3.87)\quad (2.85)\quad (2.74)\quad (2.36)\quad\quad (.40)$$

$$R^2 = .73 \qquad n = 32 \qquad s_e = 2380,$$

where Y = income; X_1 = education; X_2 = seniority; X_3 = sex; X_4 = political party affiliation, scored 0 = Democrat, 1 = independent, 2 = Republican; and the statistics are defined as above.

The estimates for the coefficients of our original variables are essentially unchanged. Also, political party affiliation is shown to have no statistically significant impact on employee's income ($t < 2$). Thus, in this particular case, regression analysis with an ordinal variable arrives at the same conclusion as the more proper regression analysis with dummy variables.

Determinants of Coal Mining Fatalities:
A Multiple Regression Example

Let us pick up the explanation of coal mining fatalities begun earlier. It is now clear that our bivariate model is incomplete. On the basis of theoretical considerations, prior research, and indicator availability, we formulate the following explanatory model:

$$Y = a_0 + b_1X_1 + b_2X_2 + b_3X_3 + e,$$

where Y = the annual coal mining fatality rate (measured as deaths per million hours worked); X_1 = the natural logarithm of the annual federal coal mining safety budget (measured in thousands of constant dollars, 1967 = 100); X_2 = the percentage of miners working underground; X_3 = a dummy variable for the President's political party, scored 0 when the President that year is Republican and 1 when Democrat; e = error.

We have already argued that the coal mining fatality rate falls in response to more vigorous safety enforcement, measured by the Bureau of Mines safety budget, X_1. Further, we contend that when the percentage of miners working underground (as opposed to strip mining) advances, the fatality rate rises. Last, we believe that the political party in the White House, X_3, makes a difference, with Democrats more likely than Republicans to take measures to reduce fatalities. Let us test these hypotheses.

Least squares yields these estimates (the data sources are those mentioned previously):

$$\hat{Y} = 1.23 - .189X_1 + .019X_2 + .046X_3$$
$$(1.75) \quad (-6.48) \quad (3.06) \quad (.84)$$

$$R^2 = .83 \qquad n = 44 \qquad s_e = .13$$

where the values in parentheses are the t ratios, the R^2 = the coefficient of multiple determination, n = 44 annual observations from 1932-1975 (the 1976 figure was not available for X_2), s_e = the standard error of estimate for Y.

These results suggest that federal safety enforcement, X_1, and the extent to which mining is carried on underground, X_2, significantly influence the fatality rate. However, the President's party, X_3, appears to have no significant impact on the fatality rate. (The t ratio for b_3 is quite far from the value of 2.) But before rejecting our hypothesis on the effect of the President's party, we should perhaps check for a multicollinearity problem. After all, it may simply be multicollinearity that is causing b_3 to fall short of statistical significance. Regressing each independent variable on the others in the equation yields $R_{x_1}^2 = .63$, $R_{x_2}^2 = .45$, $R_{x_3}^2 = .46$. The presidential party variable, X_3, when regressed on X_1 and X_2, produces an R^2 which is a long way from unity. Further, according to the R^2 for the other independent variables, they manifest at least the same degree of multicollinearity, but their regression coefficients still manage to attain statistical significance. In sum, it seems unlikely that multicollinearity is the cause of a lack of statistical significance for b_3.

We can conclude, with greater confidence, that the coal mining fatality rate is unaltered by political party changes in the White House. This causes us to revise our model specification and reestimate our equation, as follows:

$$\hat{Y} = 1.58 - .206X_1 + .017X_2$$
$$(2.80) \quad (-9.58) \quad (3.00)$$

$$R^2 = .83 \qquad n = 44 \qquad s_e = .13,$$

where the terms are defined as above.

This multiple regression model improves our explanation of the coal mining fatality rate, over our earlier bivariate regression model. The R^2, which is somewhat greater, indicates that fully 83% of the variance is being accounted for. Further, the more adequate specification has reduced the bias of the slope estimate for the safety expenditures variable, X_1. In the bivariate equation, this slope = $-.247$, which exaggerates the ability of safety budget increases to lower fatalities. Because X_2 was excluded, X_1 was permitted to account for a part of Y which should be explained by X_2. Inclusion of X_2 in our multiple regression equation shrunk the effect of safety expenditures to its proper size ($b_1 = -.206$).

Is this newly incorporated variable, the percentage of miners underground, even more important a determinant of the coal mining fatality rate than the safety budget variable? Evaluation of the beta weights provides one answer to this question. Standardizing the variables and reestimating the equation yields,

$$\hat{Y}^* = -.75X_1^* + .24X_2^*,$$

where the variables are defined as above, and standardized, as indicated by the asterisk, *. These beta weights suggest that the safety budget is a more important influence on the fatality rate than is the percentage of miners underground. In fact, a standard deviation change in the safety budget variable has about three times the impact of a comparable change in the percentage of miners underground.

What Next?

Comprehension of the material in this monograph should permit the reader to use regression analysis widely and easily. Of course, in so few pages, not everything can be treated exhaustively. There are topics which merit further study. Nonlinearity is one such topic. While relationships among social science variables are often linear, it is not uncommon for nonlinearity to occur. We spelled out the consequences of violating the linearity assumption, and provided an example of how a nonlinear relationship was straightened out by a logarithmic transformation. Other such linearizing transformations are available, whose appropriateness depends on the shape of the particular curve. Popular ones are the reciprocal,

$$Y = a_0 + b_1 \frac{1}{x} + e,$$

and the second-order polynomial.

$$Y = a_0 + b_1 X + b_2 X^2 + e.$$

(For good discussions of these and other transformations, see Kelejian and Oates, 1974, pp. 92-102, 167-175; Tufte, 1974, pp. 108-130.)

Another topic which we only touched on was the use of time-series. As noted, autocorrelation is frequently a problem in the analysis of time-series data. Take, for example, the model,

$$Y_t = a + bX_t + e_t,$$

where the subscript i has been replaced with t in order to indicate "time." Y_t = annual federal government expenditures, X_t = annual presidential budget request, e_t = error term. When we think of e_t as including omitted explanatory variables, autocorrelation appears quite likely. Suppose, for instance, that one of these omitted variables is annual gross national product (GNP); clearly, GNP from the previous year (GNP_{t-1}) is correlated with GNP from the current year (GNP_t); hence, $r_{e_t e_{t-1}} \neq 0$. This error process, in which error from the immediately prior time (e_{t-1}) is correlated with error at the present time (e_t), describes a *first-order autoregressive process*. This process can be easily detected (e.g., with the Durbin-Watson test) and corrected (e.g., with the Cochrane-Orcutt technique).[5] Other error processes are more difficult to diagnose and cure. (The problems and opportunities of time-series are introduced in Ostrom, 1978.)

In our exposition of regression, we have consciously stressed verbal interpretation rather than mathematical derivation. Given it is an introduction, such emphasis seems proper. At this point, the serious student might wish to work through the material using the calculus and matrix algebra. (For this purpose, consult the relevant sections of Kmenta, 1971, and Pindyck and Rubinfeld, 1976.)

Throughout, we have formulated *single-equation models*, either bivariate or multivariate. We could also propose *multiequation models*. These models, known technically as *simultaneous-equation models*, become important when we believe causation is two way, rather than one way. For example, a simple regression model assumes that X causes Y, but not vice versa, that is, $X \longrightarrow Y$. Perhaps, though, X causes Y, and Y causes X, that is, $X \rightleftharpoons Y$. This is a case of reciprocal causation, where we have two equations,

$$Y = a + bX + e$$

$$X = a + bY + e.$$

The temptation is to estimate each with the ordinary least squares procedure we have learned here. Unfortunately, in the face of reciprocal causation, ordinary least squares will generally produce biased parameter estimates. Therefore, we must modify our procedure, probably applying *two-stage least squares*. Reciprocal causation and the ensuing problems of estimation form the core issues of causal modeling (Asher, 1976, provides a useful treatment of this topic). Happily, a firm grasp of regression analysis will speed the student's mastery of causal modeling, as well as a host of other quantitative techniques.

NOTES

1. We are all familiar with the common example of a simple random sample, the lottery, where all the tickets are tossed and the winning ones are drawn out "at random." Stastical tests for making inferences from a sample to a population, such as the significance test, are based on a simple random sample.

2. A formula for the estimated correlation coefficient is,

$$r_{xy} = s_{xy}/s_x s_y$$

where

$$s_{xy} = \widehat{\text{covariance}}_{xy} = \frac{\Sigma(X_i - \overline{X})(Y_i - \overline{Y})}{n - 1} ,$$

and

$$s_x = \widehat{\text{standard deviation}}_x = \sqrt{\frac{\Sigma(X_i - \overline{X})^2}{n - 1}} ,$$

$$s_y = \widehat{\text{standard deviation}}_y = \sqrt{\frac{\Sigma(Y_i - \overline{Y})^2}{n - 1}} .$$

3. One might wonder why omitted explanatory variables are not simply incorporated into the equation, thus solving for autocorrelation and specification error at the same time. Lamentably, this straightforward solution is not possible when these variables are either unknown or unmeasured.

4. This example was uncovered and diagnosed entirely by my colleague, Peter Snow. He graciously allowed me to reproduce it here.

5. We might note that the Durbin-Watson test fails to reveal significant autocorrelation in the error process of our multiple regression model for coal mining fatalities.

REFERENCES

ASHER, H. B. (1976) "Causal Modeling." Sage University Paper series on Quantitative Applications in the Social Sciences, 07-003. Beverly Hills and London: Sage Pubns.

BIBBY, J. (1977) "The general linear model—a cautionary tale," pp. 35-79 in C. A. O'Muircheartaigh and Clive Payne (eds.), The Analysis of Survey Data (Vol. 2): Model Fitting. New York: Wiley.

BOHRNSTEDT, G. W. and T. M. CARTER (1971) "Robustness in Regression Analysis," pp. 118-146 in H. Costner (ed.), Sociological Methodology 1971. San Francisco: Jossey-Bass.

GERMANI, G. (1973) "El surgimiento del Peronismo: el rol de los obreros y de los migrantes internos." Desarrollo Economico: Revista de Ciencias Sociales XIII:51 (Oct.-Dec.): 435-488.

KELEJIAN, H. H. and W. E. OATES (1974) Introduction to Econometrics: Principles and Applications. New York: Harper & Row.

KERLINGER, F. N. and E. J. PEDHAZUR (1973) Multiple Regression in Behavioral Research. New York: Holt, Rinehart & Winston.

KMENTA, J. (1971) Elements of Econometrics. New York: Macmillan.

OSTROM, C. W., Jr. (1978) "Time Series Analysis: Regression Techniques." Sage University Paper series on Quantitative Applications in the Social Sciences, 07-009. Beverly Hills and London: Sage Pubns.

PINDYCK, R. S. and D. L. RUBINFELD (1976) Econometric Models and Economic Forecasts. New York: McGraw-Hill.

TUFTE, E. R. (1974) Data Analysis for Politics and Policy. Englewood Cliffs, NJ: Prentice-Hall.

MICHAEL S. LEWIS-BECK is Associate Professor of Political Science at the University of Iowa and holds a Ph.D. from the University of Michigan. He has taught quantitative methods courses at Iowa, at the Inter-University Consortium for Political and Social Research (University of Michigan), and at the European Consortium for Political Research (University of Essex, England). His substantive areas of interest are political economy and public policy. Professor Lewis-Beck has published numerous papers in scholarly journals, including the American Political Science Review *and the* American Journal of Sociology.